Success Stories in Asian Aquaculture

Sena S. De Silva · F. Brian Davy
Editors

Success Stories in Asian Aquaculture

Springer

NACA

International Development Research Centre
Ottawa • Cairo • Dakar • Montevideo • Nairobi • New Delhi • Singapore

Editors
Sena S. De Silva
Network of Aquaculture
Centres in Asia-Pacific
Bangkok
Thailand
sena.desilva@ enaca.org

F. Brian Davy
International Institute for
Sustainable Development
Ottawa
Canada
bdavy@iisd.ca

International Development Research Centre
PO Box 8500
Ottawa, ON, Canada K1G 3H9
info@idrc.ca/www.idrc.ca
eISBN 978-1-55250-461-1

ISBN 978-90-481-3085-6 ISBN 978-90-481-3087-0 (eBook)
DOI 10.1007/978-90-481-3087-0
Springer Dordrecht Heidelberg London New York

Library of Congress Control Number: 2009933584

Cover photo description: Setting of small scale cage farming activities of lobster and grouper in the Kendari Bay, S. East Sulawesi, Indonesia. Photo by S. De Silva

Printed on acid-free paper

Springer is part of Springer Science+Business Media (www.springer.com)

Foreword

World aquaculture has grown tremendously during the last 50 years, from less than a million tons in the early 1950s to 51.7 million tons with a value of US $78.8 billion in 2006. While the capture fishery production leveled off in the mid-1980s, the aquaculture sector has maintained an average annual growth rate of 8.7% worldwide since 1970. Most of the net growth in fish production over the past 20 years has come from aquaculture, especially in developing countries. Aquaculture has, in fact, been the fastest growing sector for food production, as well for rural livelihood improvement and agricultural income earning worldwide for the past two decades. It is foreseen that this trend will continue and aquaculture will be the major, if not the only, contributor to the fish supply to meet the increasing demand of the growing population in decades to come.

Asia is by far the world's leader in aquaculture production, producing more than 80% of the world's total aquaculture output in 2006. Though much has been said about Asian aquaculture, much more is yet to be understood about the reasons and factors behind such a spectacular development. It is worthy to conduct such an analysis of reasons conducive to the sustainable development of the industry, and to evaluate ways whereby these success stories of sustainable and lucrative aquaculture practices could be beneficial for other countries with aquaculture potential. It is believed by many that there is a need to document "success stories" in aquaculture, and to advise the policy makers and practitioners, as well as the general public, about the positive impact of aquaculture development in supporting rural livelihood and economic development, particularly in developing countries.

A recent FAO global study on constraints facing aquaculture found some opposition for its future development. In some cases, the cause of the opposition is a result of misinformation. The public misperception on aquaculture has occurred in some areas, leading to negative impacts on local socioeconomic development and rational use of resources. In some instances, such attitudes and perceptions toward aquaculture have influenced decision-makers, pressuring them to regulate and often to halt the expansion of aquaculture. The successful experiences and lessons learned could help to bring about policy changes, particularly in emerging aquaculture nations to pursue aquaculture development in a sustainable and an environment-friendly manner. Building awareness on success stories and highlighting the positive impact of aquaculture were considered important. The practical country case studies of successful governance models could be referred to by others in order to improve the management of the sector.

This is the initial step toward compiling stories of aquaculture successes, and the editorial team is to be congratulated for its great efforts. In approaching this difficult assignment, the team benefited from the leadership of the Network of Aquaculture Centres for Asia and the Pacific (NACA), and the support of the World Fisheries Trust (WFT) as well as the Institute for International Sustainable Development (IISD) of Canada. We hope that the team will continue its endeavor in producing other aquaculture success stories, also from other regions of the world.

Rome, Italy Jiansan Jia

Preface

We are moving into a turbulent and an uncertain era, particularly in respect of the future food needs. Given the push to sustainability, the rise in food prices, and the impending concerns around climate change and related complexity on providing the food needs for an increasing global population, it is time to address coping strategies. It is in this context that the issue on where will aquaculture development move in the future is taken up. It is our belief that one of the best options available in this regard is to address what we can learn from "success," and consequently, what types of lessons learned are likely to be most useful in guiding future aquaculture development?

We feel that the past successes of Asian aquaculture, which is the main driver in global aquaculture developments in the last three to four decades, have a great deal to contribute towards guiding its future development. Equally, we have also considered and discussed "failure" as an important learning tool, and in this book, we have sought to incorporate some of those lessons learned into this thinking as well. Overall, we felt that the flag of success was the more positive banner under which to develop this work, so we have purposefully chosen to use success as our main focus. In the following sections, we seek to explore this thinking in more detail around the meaning of success, its various audiences, some history, and the evolution of these aquatic resources systems around more to social change and outcome/impact-based thinking.

Success, ofcourse, can have many definitions. Success may refer to the achievement of one's aim or goal and financial profitability *(business)*; a person who achieves his or her goals or a level of social status, achievement of an objective/goal, or the opposite of failure.

However one chooses to define success, we feel that aquaculture has demonstrated increasing instances of success, but to the best of our knowledge, few authors have looked at these in any detail, and if done, those are in isolation only. Initially we seek to introduce the reader to how this work originated, what were the main issues guiding our thinking, and finally some thoughts on where we see this work proceeding. In the introductory section, we have tried to set the stage for the ideas and results that will follow. For instance, we outline the evolution of aquaculture and seek to position this under a set of key drivers including the environmental, changed and changing public perceptions, and moves to sustainability, recent concerns around the food "crisis," and other impending future concerns such as climate change.

Background

Following the recommendations of the Workshop on "Research Needs to Sustaining Aquaculture to 2025 and Beyond" (http://www.enaca.org/modules/news/article.php?storyid=1733&keywords=IDRC), June 2007, held in Rayong, Thailand (sponsored by IDRC Canada), the International Institute for Sustainable Development (IISD), World Fisheries Trust (WFT), and Network of Aquaculture Centers in Asia-Pacific (NACA) decided that there is an evolving demand and a clear opportunity to do more to better define future development-oriented research directions. There was agreement that aquaculture, particularly small-scale aquaculture, has always been about people and rural communities, but the examination of social organization and related issues have been relatively untouched in laying the future directions of development-oriented research in this field, for example, the new management practices evolving in Indian shrimp aquaculture (Umesh 2007). Furthermore, the questions arise as to what approaches might be tried in regions where aquaculture has a shorter (and different) history, but a significant potential scope to improve rural livelihoods, such as in Latin America, the wider Caribbean, and Africa.

Our work has involved a variety of partners from government, academia, non-government and international organizations. This group embarked on the undertaking of a comparative analysis of the lessons learned from selected "success stories" in aquaculture. It is hoped that these lessons learned from the successes dealt within this compendium will provide some guiding examples for attaining sustainability of other ongoing aquaculture practices as well as in future endeavors in the sector.

This study is intended to capture the trends and lessons learned that have driven this evolution of aquaculture augmented with comparable cases on small-scale fisheries. This material will form a part of the strategy for guiding the more detailed follow-up steps, which will hopefully set a new course for a more sustainable development of aquaculture. Our plans include the development of a series of influencing strategies to follow based on the lessons learned from this documentation and analytic phase.

In brief, our objectives were to prepare case studies and a synthesis examining the evolution and adaptation strategies in aquaculture and small-scale fisheries in the developing world.

The purpose of success stories that we have chosen are expected to:

- Highlight experiences in aquaculture development which have led to positive social change and negligible ecosystem impacts
- Provide a better understanding of the factors and approaches leading to sustainable aquaculture growth combined with positive societal change
- Show that small-scale farmers have incentives for and can act responsibly

- Provide a direction for future policy change and collective improvement to ensure sustainability of the sector
- Mitigate negative public perceptions of aquaculture

We set out to choose the list of success stories that was initially developed at the Rayong Thailand workshop through a consensus-based process by the "experienced group" assembled in Rayong. These cases were chosen to reflect activities that have impacted on livelihoods of different socioeconomic groups and have strongly contributed to aquaculture sustainability in Asia. This group promoted and endorsed the concept that more needs to be achieved to emphasize how aquaculture in Asia has been evolving particularly in terms of sustainability concerns, while attempting to meet the market opportunities and the food/nutrition gap that is developing in this region.

Implementation Strategy and Impacts

It is expected that the "success stories" project, apart from its immediate impacts on decision makers, will also lead to a variety of other activities that will bring about a wider sectoral awareness and adoption of positive lessons learned from the success stories. Such work is neither readily available, nor readily accessible at present in the published literature. Also, we hope that the project will stimulate a number of cross-level and cross-sectoral processes such as, horizontal (community to community) and vertical (community/village, district, state, national) level changes and various dialogue processes such as exchanges, farmer to farmer visits, workshops, training programs as part of the continued learning from the success program.

Bangkok, Thailand — Sena S. De Silva
Ottawa, Canada — F. Brian Davy

Reference

Umesh, N.R. 2007. Development and adoption of BMPs by self-help farmer groups. *Aquaculture Asia* XII(1), 8–11.

Acknowledgments

This compendium would not have seen the light of day if not for the workshop on "Research Needs for Sustaining Aquaculture to 2025 and Beyond," in July 2007, when the idea originated. In this regard, we are thankful to the International Development Research Centre (Canada) for financing the event. We are thankful to all those organizations that endorsed the idea of documenting selected Success Stories in Aquaculture and the lessons learned thereof was worthwhile and a much needed exercise, and came forward with financial support to this end. Accordingly, we thank the International Institute for Sustainable Development (IISD), Winnipeg, Canada, World Fisheries Trust (WFT), Victoria, Canada, and the Icelandic International Development Agency (ICEIDA) for their generous financial support. We also acknowledge the indirect support given by the Fisheries Management and Conservation Service of the Department of Fisheries and Aquaculture, of the Food and Agriculture Organization.

Needless to say that a task of this enormity and importance could not have been achieved if not for the commitment of all the authors and their respective organizations for permitting them to take part in this activity. We thank for their patience and willingness to accept and accommodate constructive criticisms and engaging in dialogues that helped improve the quality of the contributions. Our gratitude is also due to Professor Peter Edwards and Dr. Amara Yakkupitiyage (of the Asian Institute of Technology) and Dr. Tumi Tómasson (ICEIDA) for their contributions in the review process. Special thanks to Hassanai Kongkeo, Thuy T.T. Nguyen, Simon Wilkinson, Mala Amarasinghe, and Nongluk Pituktammanat of NACA for going beyond their normal duties and contributing to the process from its initiation.

Last but not least, our appreciation and gratitude to the small-scale, rural aqua farmers of Asia for making this story a possibility and for continuing to contribute to feeding the ever increasing hunger in the world.

Bangkok, Thailand — Sena S. De Silva
Ottawa, Canada — F. Brian Davy

Contents

Chapter 1
Aquaculture Successes in Asia: Contributing to Sustained Development and Poverty Alleviation

Sena S. De Silva and F. Brian Davy

Abstract Aquaculture, though considered to have over a 2500 year history, was mostly practiced as an art. It began to be transformed into a modern science in the second half of the 20th century. Within a period of 25 years or more, it had begun to impress upon as a major food production sector, having recorded an annual average growth rate of nearly 8% in the last two decades, as often purported to be the fastest growing primary production sector. Currently, aquaculture accounts for 50% of the global food fish consumption.

The sector has been and continues to be predominant in developing countries, particularly in Asia, which accounts for more than 85% of the global production. Asian aquaculture by and large is a small scale farming activity, where most practices are family-owned, managed, and operated. The sector has provided direct and indirect livelihood means to millions, a significant proportion of which is rural, and for some Asian nations, it is a main source of foreign exchange earnings. Furthermore, it has contributed to food security and poverty alleviation, and is considered to be a successful primary food sector globally.

1.1 Introduction

Fish/aquatic food organisms have been inextricably linked with human life over millennia. Indeed, it is even suggested that the prime impacting factor on the evolution of the human brain, which has made us what we are today, is linked to our early ancestors depending on aquatic food sources as the main form of nourishment that provided ample quantities of n-3 and n-6 series highly unsaturated fatty acids. *Homo sapiens* in the recent history went through agricultural and industrial revolutions, which gradually impacted on our life modalities including food sources.

S.S. De Silva (✉) and F.B. Davy
Network of Aquacultures Centres in Asia-Pacific, PO Box 1040, Kasetsart Campus, Bangkok, 10903, Thailand
e-mail: sena.desilva@enaca.org

S.S. De Silva and F.B. Davy (eds.), *Success Stories in Asian Aquaculture*,

Science began to have a profound influence on our lifestyles and gradually the life expectancy increased and the ability to combat epidemics of various forms and sorts developed and began to impact on the rate of growth of human population. In the early years of the postindustrial revolution period, the early philosophers such as Malthus held a grim view of the sustenance of the growing human population, and suggested that gains in food production will not be able to keep up with the latter (Malthus 1985). He suggested that there will be a constant tendency in all animated life to increase beyond the nourishment prepared for it, which would result in an inevitable number of "positive checks" (e.g., starvation and deaths), occurring when mouths outnumbered food production.

Impacts of science on our life have changed the above view. Indeed, some economists (Boserup 1981; Simon 1981) have suggested that man is a resource, and the greater the population, the more likely it is that invention and innovation will flow and food will not become a limiting factor. Of course the latter position is not universally accepted either, especially in the context of the notion that there is a limit to the capacity in the biosphere, in that it is essentially a closed system, and continuing sustainable gains in food production cannot be taken for granted.

In this opening chapter, an attempt is made to set the scene for the treatise as a whole. Accordingly, the importance of aquaculture in the current context of population growth, food demand, poverty and malnourishment prevailing globally, and its role in contributing to human food basket is brought to focus. The growth of the aquaculture sector is traced, and the public perceptions associated with aquaculture development in respect of its sustainability are discussed. The importance of public perceptions in current development activities and on aquaculture is considered, and finally an attempt is made to answer the simple question whether aquaculture is a success.

1.2 Contemporary Situation

The human population has grown from 1.5 to 6.4 billion from 1900 till now, and is predicted to increase to 9 billion by the year 2050, barring major calamities that could occur. Not surprisingly, the fact remains that malnourishment, defined as human beings' daily calorie intake is less than 2,200 KCal, is probably one of the challenges if not the biggest challenge facing the globe, with an estimated 840 million being in a state of malnourishment (UNWFP 2005). The situation is even more burgeoning as nearly 80–85% of these malnourished people live in the developing world, accounting for 16% of the world's total burden of disabling illness and premature death, measured in disability-adjusted life-years (Murray and Lopez 1996). Fish, however, is not primarily a source of calories. On the other hand, fish accounts approximately 20% of the global animal protein intake, the contribution being even higher in developing countries, particularly those of Asia.

Over the last half century or so, fish are also becoming an increasingly important traded commodity, bypassing many traditionally traded commodities (Kurien 2005). Moreover, in a significant number of Asian countries, the contribution of aquaculture to the national GDP has superseded that from capture fisheries, indicating the growing importance of aquaculture to those countries (Table 1.1).

Recently McMichael (2001) recognized three consecutive eras over the past and coming centuries show the changing balance between cereal-grain production and population growth (Fig. 1.1). Furthermore, he summarized the contemporary situation under seven headings, of which the pertinent points are as follows:

Table 1.1 The relative contribution from capture fisheries and aquaculture to the GDP in some selected Asian countries and Chile, S. America

Country	Capture	Aquaculture
Bangladesh[a]	1.884	2.688
PR China[a]	1.132	2.618
Indonesia[a]	2.350	1.662
Lao PDR[a]	1.432	5.775
Malaysia[a]	1.128	0.366
Philippines[a]	2.184	2.633
Thailand[a]	2.044	2.071
Vietnam[b]	3.702	4.00
Chile[c]	2.17	2.63

[a]From RAP (2004)
[b]Vietnam Net Bridge
[c]From De Silva and Soto (2009)

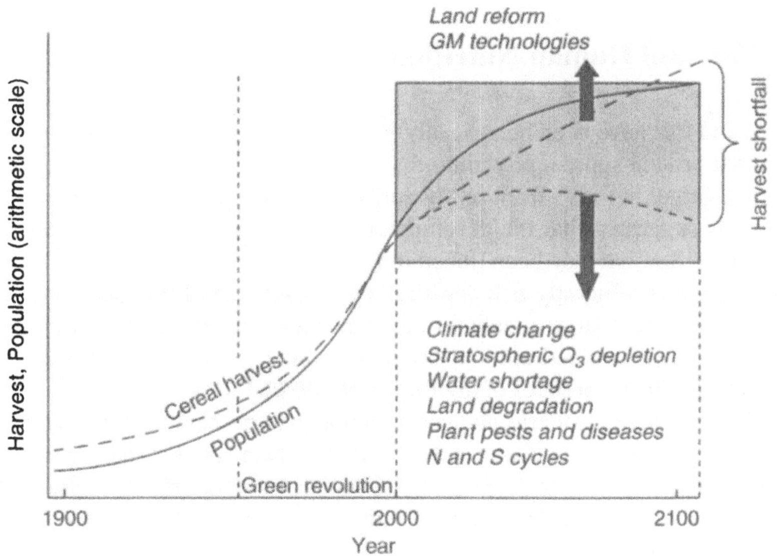

Fig. 1.1 Schematic representation of three consecutive eras over the past and coming centuries, showing the changing balance between cereal-grain production (dashed lines) and world population size. Over the coming century, there will be tension between yield-enhancing science and policies and yield-diminishing environmental forces. *GM* genetically-modified

(1) the proportion of the world population that is hungry and malnourished is slowly declining. However, because of the continuing growth in population, the absolute number of hungry and malnourished people is not obviously declining and one-quarter of them are aged below 5 years
(2) *per capita* food production has increased over the past four decades, but those gains in yield were achieved via intensive inputs of energy, fertilizer, and water, and at the expense of soil fertility and groundwater stocks
(3) an estimated one-third of the world's arable land is significantly degraded, a significant proportion of it (20%) occurring during the 1980s and 1990s
(4) *per capita* grain production (which accounts for two-thirds of the world food energy) has plateaued since the mid-1980s
(5) the annual harvest from wild fisheries peaked at about 100×10^6 t per year in the 1970s, and has subsequently declined by about 20%
(6) aquaculture now accounts for approximately one quarter of the world's total fish and shellfish production, and
(7) the promise of genetically-modified food species, while potentially great, is subject to resolution of concerns about unexpected genetic and ecological consequences

Based on the above, McMichael proceeded to address issues of the impact of climate change on food production, as well as the associated scenarios on biofuels production, for example.

1.3 Fish and Human Nutrition

Humans and fish have been inextricably linked for millennia, not only as an important animal protein source, providing many millions of livelihood means and food security at large, but also from an evolutionary viewpoint. Indeed, one school of thought has suggested that the development of the human brain, and hence, what humans are today, has also been linked to food sources rich in n-3 (DHA, EPA) and n-6 (AA) PUFAs – literally fish constituting a major part of the diet of our ancestors. In this regard, a large quantum of evidence has been brought forward to show that *Homo sapiens* evolved not in a savannah habitat, but in a habitat that was rich in fish and shellfish resources (Crawford et al. 1999).

The fish production patterns and consumption patterns have changed over the last 30–40 years, with both production and consumption being predominant in developing countries (Delgado et al. 2003). Fish and all aquatic products are easily digested, and though perishable, are easily processed into various forms avoiding wastage. Most importantly, fish constitutes one of the main animal protein sources of the developing world, containing all essential amino acids, thereby providing an affordable nutrient source to most rural, impoverished communities. Fish also provides an excellent source of essential fatty acids, the highly unsaturated acids of the n-3 and n-6 series [e.g., DHA – docosahexaenoic acid, 22(6n-3); EPA – eicasopaentonic acid, 20(5n-3);

AA – arachidonic acid, 22(4n-6)], though the amounts of the specific fatty acids present in fish differ markedly between species, and in general, between those of marine and freshwater origin. Fish also provides essential micronutrients in the form of vitamins, mineral (e.g., best sources of iodine and selenium), and some co-enzymes (CoQ 10), among others. Increasing quantum of evidence is becoming available on the health benefits of fish consumption, with clear evidence being brought forward with regard to its impacts on common diseases such as cardio vascular related ones (de Deckere et al. 1998; Horrocks and Yeo 1999; Connor 2000; Ruxton et al. 2005, among others).

It is in this respect that there is an increase in fish consumption in the developed world, whereas in the developing world, in all probability, the driving forces with regard to increased consumption are its affordability and availability. There is clear evidence that both in the developing and developed world, fish consumption is on the increase and more so in the former (Delgado et al. 2003).

1.4 Traditional Fish Food Supplies

Traditionally, the great bulk of the food fish supplies (conservatively estimated at about 85–90%) were of marine origin, and it still is, but its share is declining. Historical developments of the industrial fisheries, which essentially were a post-World War 2 development, have been aptly documented in the past (FAO 2007). Importantly, now for over four decades, about 25% of the industrial fish landings are reduced into fish meal and fish oil, currently approximating 25 million tons per year, and form the basis for the animal feed industry, including some cultured aquatic species/species groups, such as shrimp and carnivorous marine finfish in the main (Fig. 1.2).

Until the late 1960s, it was thought the seas were bountiful and harbored unlimited amount of fish resources. This notion is now known to be proven erroneous. The intensification of fishing has resulted in the depletion of stocks, perhaps some beyond recovery, and the number of potential stocks that could be added to the fish basket is known to be meager. It is almost universally accepted that the average maximum fish production that could be obtained from the oceans is around 100×10^6 tons per year.

1.5 Fish Food Needs

Although the production from the marine capture fisheries has plateaued, the demand for fish has grown over the years, resulting from an increased global population, exacerbated by increased consumption in some nations/regions. The estimates of food fish needs for the future, even at the current rate of consumption, are high. Table 1.2 summarizes the fish food needs upto 2020 based on the average of different estimations that are available. In summary, the world will need an extra $40–60 \times 10^6$ tons of food fish by 2020.

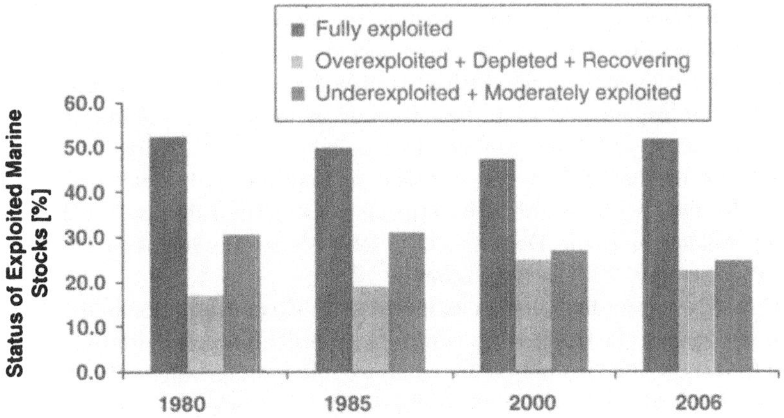

Fig. 1.2 The status of exploited marine fish stocks (based on FAO 2007)

Table 1.2 Estimation of fish food needs by 2020 based on the current per caput consumption rates

Continent	Population (× 1000)[a]			Fish supply/demand		
	2005	2020	Increase (%)	Per caput 2001 (kg)[b]	Current (tons)[c]	Demand 2020 (tons)[d]
Africa	905,936	1,228,276	35.6	7.8	7,066,301	9,580,553
Asia (e.g., China)	2,589,571	3,129,852	20.9	14.1	36,512,951	44,130,913
Europe	728,389	714,959	−1.8	19.8	14,422,102	14,156,188
L. America and Caribbean	561,346	666,955	18.8	8.8	4,939,845	5,869,204
N. America	330,608	375,000	13.4	17.3	5,719,518	6,487,500
Oceania	33,056	38,909	17.7	23	760,288	894,907
China	1,315,844	1,423,939	8.2	25.6	33,685,606	36,452,838
World	6,464,750	7,577,889	17.2	16.3	105,375,425	123,519,591

[a]Source: UN
[b]Source: FAO
[c]2005 population × 2001 per capita supply
[d]2020 population × 2001 per capita supply

1.6 Aquaculture

Aquaculture is purported to be an age old practice that commenced in China. However, its significance to the contribution to "human food basket" is of only three to four decades old. There is universal acceptance that aquaculture has matured to be an important contributor to meet human food fish demands, and is often mooted as the fastest growing primary production sector.

1.6.1 Importance in Narrowing the Supply and Demand Gap

In the wake of the plateauing of the traditional fish supplies, it is generally accepted that the shortfall in food fish needs has to come from aquaculture. Aquaculture within the short span of three to four decades has continued to increase its importance in the food fish production sector, and is estimated to account for 50% of the global food fish consumption (Fig. 1.3), and tantamount to approximately 35% of the global fish production/availability.

1.6.2 Key Features of the Aquaculture Sector

The great bulk of aquaculture production occurs in Asia, and within Asia, in China. Freshwater finfish culture contributes most to overall production (Figs. 1.4 and 1.5). In general, the great bulk of aquaculture production is based on commodities that command a farm gate price of less than US$ 2.00 kg^{-1}, and all in all over the years, the price of aquaculture produce has retained static and/or declined marginally, in stark contrast to other food commodities. Of the major cultured commodities from a unit (kg) value viewpoint, the most important cultured commodities are the salmonids and shrimp in temperate and tropical regions, respectively.

In Asia, the epicenter of aquaculture production, the traditional practices tend to be largely small scale operations, often farmer-owned and managed, and often clustered in an area that is conducive to aquaculture. Aquaculture practices are integrated with other primary production practices, particularly in Asia, such as rice farming.

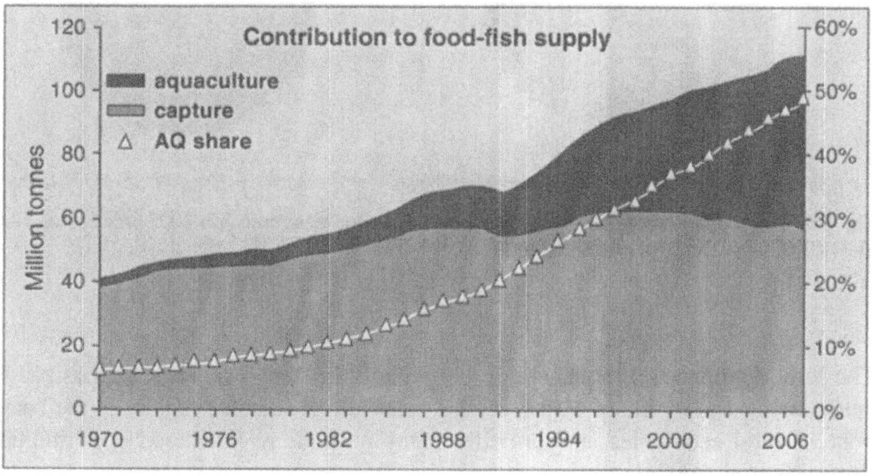

Fig. 1.3 Contribution from capture fisheries and aquaculture to fish production and the percent contribution of the latter to consumption (from Subasinghe et al., 2009)

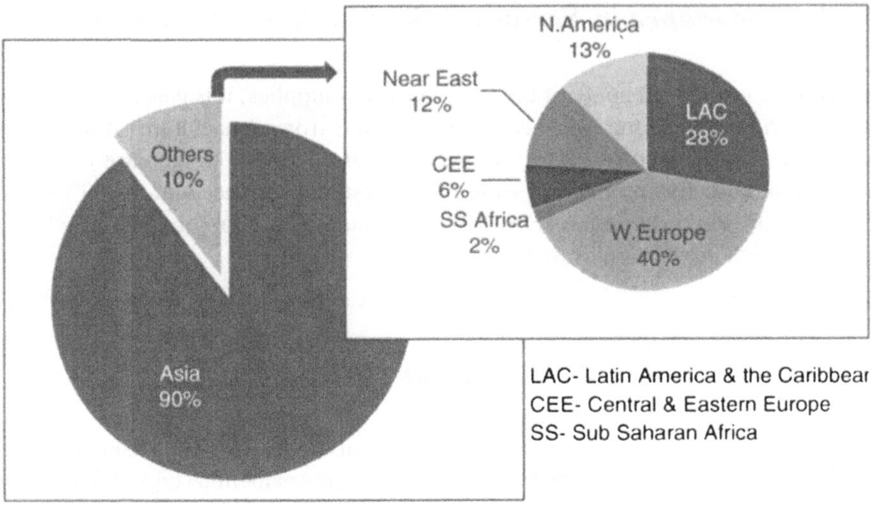

Fig. 1.4 Aquaculture production (2005) in relation to the geographic areas (from De Silva and Soto, 2009)

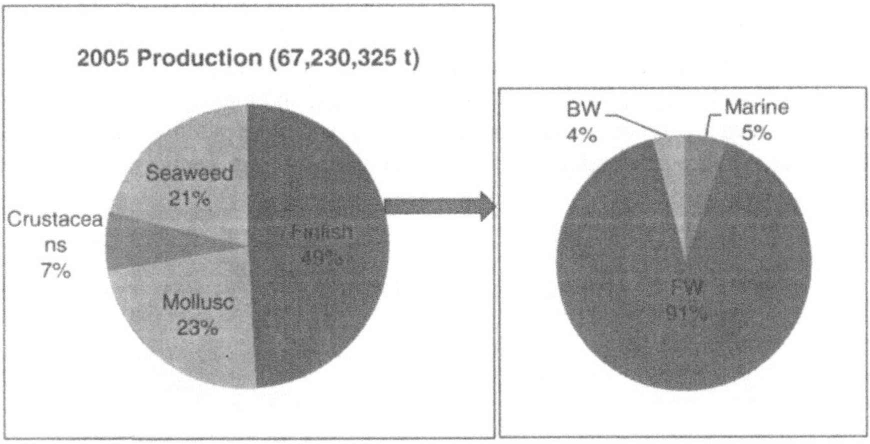

Fig. 1.5 The proportion of the major aquaculture commodities (based on 2005 production, FAO Stats 2007; from De Silva and Soto, 2009)

The rural dominance of aquaculture, particularly in Asia, has been emphasized in detail earlier and by many. In this regard, aquaculture in some regions is the backbone of rural economies, and provides many millions of livelihood opportunities (Edwards et al. 2002).

Needless to say that in the last decade or so, many rural aquaculture farming practices have changed, the changes being primarily in response to market demands, traditional and new. Notable examples in this context are the tra catfish (*Pangasianodon hypophthalmus*) and rohu (*Labeo rohita*) farming sectors of the Mekong, Vietnam and Ayeyarwaddy, Myanmar deltas, respectively. These have remained rural intensive and have flourished to cater to export niche markets, and gone on to generate thousands of additional livelihood opportunities, particularly for women, in the associated processing sectors (Aye et al. 2007). Inevitably such developments have also highlighted on the difficulties of defining small scale aquaculture, as explicitly as one could. For example, a shrimp farming practice in 1 ha could result in an optimal annual yield of approximately 8–10 tons, as opposed to a catfish farming practice in the Mekong delta that could yield 400–600 tons.

1.6.3 Growth Phases in Aquaculture

As pointed out earlier, the aquaculture sector has grown fast, averaging almost 7–8 percent annually, over the last two decades, significantly higher than any other primary production sector (De Silva 2001). It will be naïve to expect that an almost exponential rate of growth could be maintained indefinitely, and evidence is coming forth in that globally the relative rate of growth of the sector, expressed as a percentage of the global aquaculture production during the defined period, is declining (Fig. 1.6; also see De Silva and Hasan 2007). In Fig. 1.6, the rate of growth of human population is superimposed, and some degree of similarity in the trends is apparent. Indeed, if the trends in absolute increase in aquaculture production and growth in human numbers are considered, the trends appear to be somewhat comparable (Fig. 1.7). Ideally, it would be of further interest to consider the changes in the growth rate of human populations of countries in which fish eating habit is predominant with the corresponding changes in aquaculture growth.

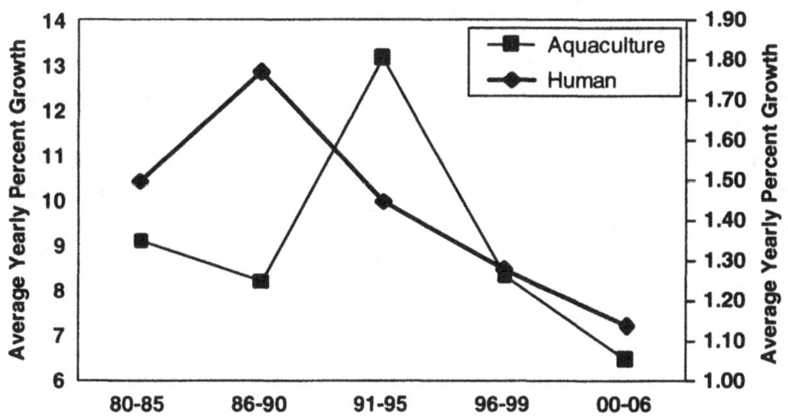

Fig. 1.6 Trends in aquaculture production, based on percent average change in production per year in each of the 5-year periods, 1985–2005 together with that of human population growth

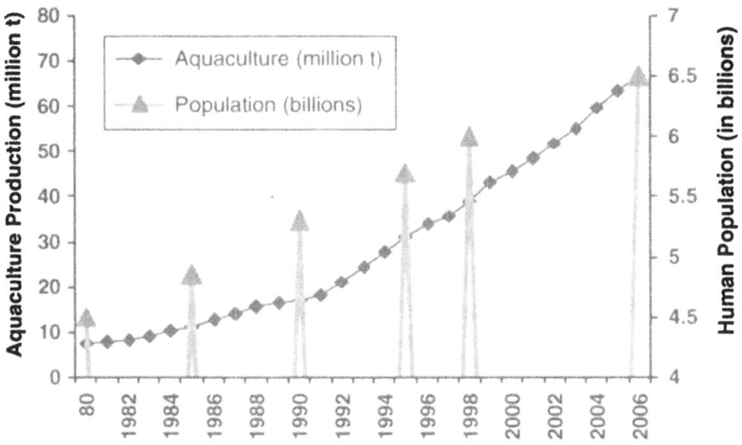

Fig. 1.7 The trend in aquaculture production (FAO 2008) over the years and the corresponding human population numbers (data source: http://geography.about.com/od/obtainpoulationdata/a/worldpoulation/htm)

A detailed analysis of some of the above trends has also been provided earlier (De Silva 2001), when it was pointed out that although there is a global decline in the rate of growth, in some regions such as in Latin America and Africa, the rate is on the increase, albeit the overall contribution to global aquaculture production is still relatively small from these continents. However, this also indicates that there is a high potential for aquaculture growth in these regions. Overall, therefore, although the rate of growth in the aquaculture sector is decreasing, the absolute production continues to increase. What is crucial is to recognize that there will be a limit to this growth, and sustaining the optimal production when it reaches that point.

Over the last four to five decades, the aquaculture sector has gone through some notable phases. The latter phases could be related to changes in global aspirations in respect of development goals, commencing with the worldwide acceptance of the Bruntland Report (UNEP 1987). The progress of the sector, therefore, had to accede to public well-being in a holistic sense, and had to work toward a more prudent use of primary resources, such as land and water, and be much more conscious of environmental impacts arising from the sector's developments. A schematic representation of the conceived phases of aquaculture growth is presented in a schematic form in Fig. 1.8.

The question, therefore, arises that whether the sector is capable of complying with global aspirations and what adjustments are needed to meet such a compliance. The present attempt to document the "success stories" in aquaculture is one such manner of demonstrating to a global audience that aquaculture could comply with global aspirations and continue to contribute significantly to the human fish food basket.

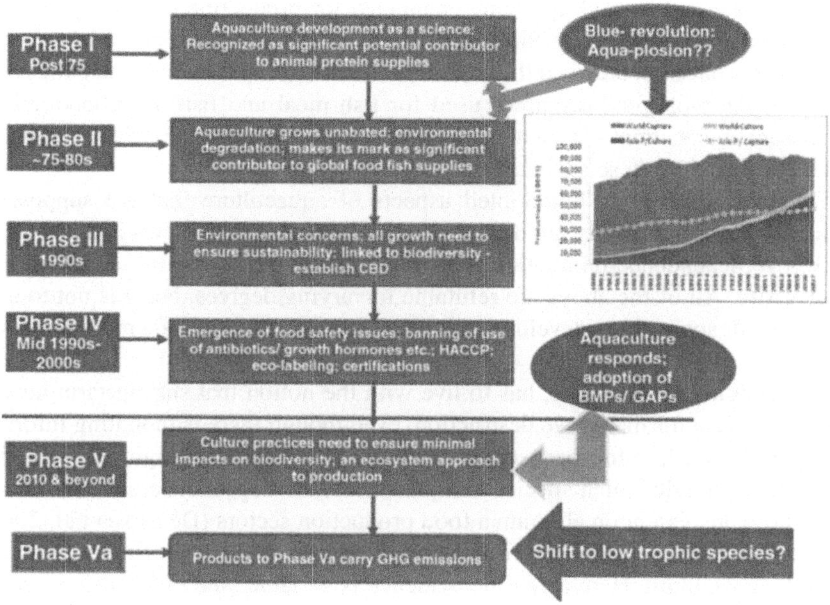

Fig. 1.8 A schematic representation of growth phases of modern aquaculture

1.6.4 Public Perceptions on Aquaculture

Aquaculture is a relatively new primary production sector from the viewpoint of its significant and important contribution to the human food basket. However, this increasing importance happened to occur when the world as a whole became conscious and realized that all developments, be it in the primary production sector or elsewhere, have to be sustainable and environmentally minimally perturbing, and have to comply with increasing food quality and nutritional needs. Consequently, the sector has had to encounter much more stringent public, scrutiny and "policing." The current scenario is such it is not only important that present day food quality demands are met, but equal importance is being laid on the manner it was produced, including societal responsibility and equity.

Many public groups reckon that aquaculture is unsustainable and environmentally degrading (Naylor et al. 1998, 2000; Aldhous 2004; Allsopp et al. 2008). However, most of these perceptions revolve around shrimp and salmonid aquaculture, two practices that contribute less than 10% to overall global production and in value approximately 16%, and more often than not fail to consider its contribution as a food source, an avenue for providing millions of livelihood opportunities and contributing to the national and global trade. The major objections to aquaculture development is based on excessive use of fish meal and fish oil in the feeds for shrimp and other cultured marine carnivorous species/species groups. It has been shown that very significant proportions of the primary resource used for reduction into fish meal and fish oil are being channeled for production of animals that are not human food sources (De Silva and Turchini 2008). What is perturbing is that such aspects have not received the attention of these groups who advocate the channeling of the biological resources used for fish meal and fish oil production for direct human consumption; on many an occasion, a level playing field has not been lacking in many of these advocacies.

The most commonly highlighted aspects of aquaculture that are supposedly impacting its sustainability are dependence on fish meal and fish oil in aquaculture feeds, overdependence on exotic and/or alien species, and negative influences on biodiversity. All of the above are refutable to varying degrees, but it is not to conclude that all aquaculture developments have been sustainable and environmentally friendly.

Aquaculture, even to date, has to live with the notion that shrimp farming was the prime cause for mangrove destruction, even though there is mounting information that shrimp farming per se accounted for less than 5% of global mangrove destruction. Equally, alien species in aquaculture, the dependence on which is not different to any extent on all human food production sectors (De Silva et al. 2008), are often highlighted as an environmentally destructive element primarily induced through aquaculture. Here again the evidence is variable, and on occasions, refutable and in more often than not explicit evidence on cause and effect is lacking.

Irrespective of the correctness and/or the authenticity, and/or the biases of the varying range of arguments, it is incumbent on all aquaculture developments to take notice of the public concerns and act accordingly – aquaculture developers cannot be complacent. The onus is on the sector to demonstrate to the public and policy makers equally and convincingly that aquaculture developments are sustainable and it could proceed in a socially and environmentally responsible manner, within the societal guidelines, and still continue to be a dominant and important provider to human food basket, millions of livelihoods, and income generation, and in turn, contribute to food security and global poverty alleviation.

1.7 Is Aquaculture a Success?

Success can have many definitions, and is equally interpreted in many ways. Success may refer to the achievement of one's aim or goal; (*business*) financial profitability, a person who achieves his or her goals or a level of social status, achievement of an objective/goal; the opposite of failure. On the other hand, what is considered a success for one person, a community, a nation or region, and/or a combination thereof in respect of a policy and achievement, a development may not be considered a success by others. Indeed, even a consensual success at one point could be treated as failure by others, downstream, and is of particular relevance to aquatic resources-related developments.

If one was to assess the success of aquaculture from a purely food production, providing employment and/or livelihoods opportunities, rural development and financial gain viewpoints, it has to be conceded that aquaculture has been a success thus far. However, as pointed out in the foregoing section, this success as well as its long-term sustainability is questioned by some, and in more instances than one, the arguments are based on shrimp and carnivorous marine finfish culture of salmonids. It is in the above context that the other chapters in this treatise set out to demonstrate the successes of specific aquaculture practices, which have collectively contributed to achieving significance as a food production system.

Development and success go hand in hand with public policy as well as with attitudes. Accordingly, there is a need to address decision makers in the wake of increasing use, and importance of coastal and inland waters in a timely manner for a number of different reasons. Stabilization of wild fish catches has led to an increased dependence on aquatic farming, which is being seen as the next best alternative to meet the shortfall in demand for aquatic foods. It is in this context that lessons learned from case studies presented here provide a basis for wider dissemination and lesson learning. In particular for newly emerging aquaculture nations, these provide an important basis of lessons, including the steps that are needed to facilitate developments and help guide future development, rather than on ad hoc basis. Equally, the lessons learned will also be helpful toward attaining sustainability of some ongoing practices, both in Asia and elsewhere. There is an overall need for decision makers to be made aware that ad hoc developments are not the solution to the problem, but developments based on lessons learned from the past successful experiences, suitably modified and adapted, are a more effective way to proceed. It is regrettable that such "show casing" under one banner is hard to find. The present initiative aims to fill this gap in the interim and then carry forward the strategy in the long term in respect of aquaculture.

References

Aldhous, P. 2004. Fish farms still ravage the sea. Sustainable aquaculture takes one step forward, two steps back. *Nature Online*, 17 February 2004. doi: 10.1038/news040216-10.

Allsopp, M., P. Johnston, and D. Santillo. 2008. *Challenging the aquaculture industry on sustainability*, 22. Amsterdam: Greenpeace.

Aye Khin, Maung, Khin Ko Lay, Hla Win, and S.S. De Silva. 2007. A new freshwater aquaculture practice that has successfully targeted a niche export market with major positive societal impacts: Myanmar. *Aquaculture Asia* XII(4): 22–26.

Boserup, E. 1981. *Population and technological change*. Chicago, IL: University of Chicago Press.

Connor, W.E. 2000. Importance of n-3 fatty acids in health and disease. *American Journal of Clinical Nutrition* 71: 171S–175S.

Crawford, M.A., M. Bloom, C.L. Broadhurst, W.F. Schmidt, S.C. Cunnane, C. Galli, K. Gehbremeskel, F. Linseisen, J. Lloyd-Smith, and J. Parkington. 1999. Evidence for the unique function of docosahexaenoic acid during the evolution of the modern hominid brain. *Lipids* 34: S39–S47.

de Deckere, E., O. Korver, P.M. Verschuren, and M.B. Katan. 1998. Health aspects of fish and n-3 polyunsaturated fatty acids from plant and marine origin. *European Journal of Clinical Nutrition* 52: 749–753.

De Silva, S.S. 2001. A global perspective of aquaculture in the new millennium. In *Aquaculture in the Third Millennium*, eds. Subasinghe, R.P., P. Bueno, M.J. Phillips, C. Hough, S.E. McGladdery, and R. Arthur, 431–459. Technical Proceedings of the Conference on Aquaculture in the Third Millennium, Bangkok, Thailand, 20–25 February. Bangkok: NACA.

De Silva, S.S., and M.R. Hasan. 2007. Feeds and fertilizers: the key to long-term sustainability of Asian aquaculture. In *Study and analysis of feeds and fertilizers for sustainable aquaculture development*, eds. Hasan, M.R., T. Hecht, S.S. De Silva, and A.G.J. Tacon, 19–48. *FAO Fisheries Technical Paper* 497. Rome: FAO.

De Silva, S.S. and G.M. Turchini. 2008. Towards understanding the impacts of the pet food industry on world fish and seafood supplies. *Journal of Agriculture and Environmental Ethics* 21: 459–467.

De Silva, S.S., Nguyen, T.T. Thuy, G.M. Turchini, U.S. Amarasinghe, and N.W. Abery. 2008. Alien species in aquaculture and biodiversity: a paradox in food production. *Ambio* 37:24–28.

De Silva SS, and D. Soto, 2009. Climate change and aquaclture: Potential impacts, adaptation and mitigation. *FAO Fisheries Technical Paper*, 530: 137–215.

Delgado, C.L., N. Wada, M.W. Rosegrant, S. Meijer, and M. Ahmed. 2003. *Fish to 2020. Supply and Demand in Changing Global Market*, 226. Washington, DC: International Food Policy Research Institute.

Edwards, P., D. Little, and H. Demaine. 2002. *Rural aquaculture*, 357. UK: CABI Publishing, CAB International.

FAO. 2007. *The state of world fisheries and aquaculture*. Rome: FAO.

FAO. 2008. FishStat: Universal software for fishery statistical time series V. 2.3. Data and Statistics Unit, FAO, Rome. Available at URL: http://www.fao.org/fishery/statistics/software/fishstat/en

Horrocks, L.A. and Y.K. Yeo. 1999. Health benefits of docosahexaenoic acid (DHA). *Pharmacological Research* 40: 211–225.

Kurien, J. 2005. Responsible fish trade and food security, 102. *FAO Fisheries Technical Paper* 456.

Malthus, T. 1985. *An essay on the principle of population (1798)*. London: Penguin Classics (reprint).

McMichael, A.J. 2001. Impact of climatic and other environmental changes on food production and population health in the coming decades. *Proceedings of the Nutrition Society* 60: 195–201.

Murray, C.J.L. and A.D. Lopez. 1996. *The global burden of disease. A comprehensive assessment of mortality and disability from diseases, injuries, and risk factors in 1990 and projected to 2020*. Boston, MA: Harvard University Press.

Naylor, R.L., R.J. Goldburg, H. Mooney, M. Beveridge, J. Clay, C. Folke, N. Kautsky, J. Lubchenco, J. Primavera, and M. Williams. 1998. Nature's subsidies to shrimp and salmon farming. *Science* 282: 883–884.

Naylor, R.L., R.J. Goldburg, J. Primavera, N. Kautsky, M. Beveridge, J. Clay, C. Folke, J. Lubchenco, H. Mooney, and M. Troell. 2000. Effect of aquaculture on world fish supplies. *Nature* 405: 1097–1024.

RAP. 2004. Status and potential of fisheries and aquaculture in Asia and the Pacific. RAP Publication 2004/25, Regional Office for Asia and the Pacific, Bangkok.

Ruxton, C.H.S., P.C. Calder, S.C. Reed, and M.J.A. Simpson. 2005. The impact of long-chain n-3 polyunsaturated fatty acids on human health. *Nutrition Research Reviews* 18: 113–129.

Simon, J. 1981. *The ultimate resource*. Princeton, NJ: Princeton University Press.

Subasinghe R, Soto D, Jia J, 2009. Global aquaculture and its role in sustainable development. *Reviews in Aquaculture*, 1, 2–9.

UNEP. 1987. *Our common future*. The World Commission on Environment and Development, Commission for the Future. Geneva, Switzerland.

UNWFP. 2005. The state of food insecurity in the world. Food and Agriculture Organization, United Nations World Food Programme.

Chapter 2
Recent Developments in Rice-Fish Culture in China: A Holistic Approach for Livelihood Improvement in Rural Areas

Miao Weimin

Abstract Rice field-fish culture, also popularly referred to as rice cum fish culture, is a traditional integrated fish-rice production system. The earliest practices can be traced back to more than 2,000 years ago. China is the largest producer of fish and rice in the world. Rice-fish culture has achieved significant development in China in the past three decades, in spite of the major socioeconomic changes that have occurred during this period. There are some 1.55 million ha of rice-fish culture in China now, which produces approximately 1.16 million tons of fish products (2007), in addition to about 11 million tons of high quality rice. Fish production from rice–fish culture has increased by 13-fold during the last two decades in China. Rice-fish culture is now one of the most important aquaculture systems in China. While making significant contribution to rural livelihood and food security, development of rice-fish culture is an important approach for environment friendly holistic rural development, and epitomizes an ecosystems approach to aquaculture.

Rice-fish culture in China utilizes a range of production systems and practices, but all contribute to eco-environmental benefits and sustainable development. Many factors have contributed to these developments, but equally and still, there are challenges that need to be addressed for up-scaling these production systems and practices.

It is estimated that the area under rice cultivation in Asia approximates 140.3 million ha, accounting for 89.4% of the world total. The potential for development of rice-fish culture is very high in the region. The successful experiences and lessons of rice-fish culture development drawn from China can be a good reference for sustainable rice-fish culture development in the region as well as other parts of the world, thereby contributing further to food security and poverty alleviation.

M. Weimin (✉)
FAO Regional Office for Asia and the Pacific, Maliwan Mansion, 39 Phra Atit Road, Bangkok, 10200, Thailand
e-mail: Weimin.Miao@fao.org

S.S. De Silva and F.B. Davy (eds.), *Success Stories in Asian Aquaculture*,

2.1 Introduction

China has a very long history of aquaculture. The earliest aquaculture practices started some 2,500–3,000 year ago in China. Chinese aquaculture production has ranked top most in the world for nearly two decades due to fast and steady development in the past three decades. Aquaculture has been providing a very important source of animal protein for Chinese people, currently estimated at 25 kg/capita of aquatic products, accounting for about 70% of the total available aquatic products, and more than 20% of total animal food supply for the Chinese people. Moreover, aquaculture plays significant role in rural livelihoods and national economic development. As an important subsector of agriculture, aquaculture accounts for approximately 2% of the national GDP, and its contribution to the latter exceeded that from the capture fisheries nearly a decade back (National Statistical Bureau 2008). All these have occurred in China currently in spite of the fast pace of industrialization and associated changes in the socioeconomic conditions.

Compared with other subsectors of agriculture, aquaculture has been a more dynamic food production subsector in China. It is characterized by much diversified farming systems and practices and diverse range of organisms utilized. The subsector has changed and restructured to adapt to the changes in socioeconomic and environmental conditions in China during the last three decades. Some aquaculture systems and practices are effective in adapting to such changes and still continue to play an important role in improving rural livelihoods and contribute to food security. Fish culture in rice fields is one such farming system.

In the tropics, fish has been exploited over millennia as a second crop from a range of aquatic habitats associated with rice fields, and has led to the development of rice–fish culture in the Chinese and Indian regions. However, use of high levels of fertilizer and pesticides and introduction of agribusiness techniques had negative effects on the development of this form of fish culture, especially in the 1960s. Recent trends of application of integrated pest management and eco-friendly agronomic practices resurrected once forgotten rice–fish culture in the Philippines and Thailand (dela Cruz et al. 1992), China (MacKay 1995), Hungary (Szito and Jones 1997), and Egypt (Shehadeh and Feidi 1996; El Gamel 1997). Worldwide coverage of literature on rice–fish culture is reported by Fernando (1993, 1996), Halwart (1998), Fernando and Halwart (2000, 2001), and Fernando et al. (2005).

Fish culture in rice fields is an integrated aquaculture system with a long history in China. In the past three decades, it has developed into a more holistic manner and adapted to significant socioeconomic and environmental changes in China. It has retained its importance in inland aquaculture sector and rural development with innovative changes. It is now one of the most environmentally friendly and socio-economically viable aquaculture systems in China.

Asia is the most important region for production of both rice and fish in the world. Some 140 million ha of rice fields, nearly 90% of the world total, occur in Asia (FAO 2008). Development of rice-fish culture can make substantial contributions to rural livelihood improvement and food security in Asia (Li 1988; Pullin and Shehadeh 1980; de la Cruz et al. 1992). The objective of this presentation is to review and evaluate the successful development of rice-fish culture in China and provide valuable experiences and lessons for comparable developments elsewhere.

2.2 Development of Fish Culture in Rice Fields in China

2.2.1 Historical Aspects

China is a country with long history of rice plantation and fish culture. Areas producing both fish and rice were considered as heaven on the earth in the ancient times. Both rice and fish play extremely important roles in the life of Chinese people traditionally and presently, and are among the most important food items of food basket of Chinese people. Rice is one of the most important grain crops in China. There are some 25 million ha of rice fields in China presently (Meng and Wu 2002).

Integrated fish and crop farming is a traditional food production system in China, and fish culture in rice fields started some 2,000 years ago in central China, such as in the Sichuan basin and Hanzhong areas. Such practices were documented in ancient writings and also depicted in ancient pottery unearthed from the tombs of East Han Dynasty with designs of rice fields with fish (Zeng and Wang 2006).

Traditional fish culture in rice fields remained as a simple spontaneous practice until the late 1950s when integrated fish farming was promoted by the government, when the total area of fish culture in rice fields reached nearly 700,000 ha in by 1959. Limited by technical know-how and management skills, both the unit production and economic efficiency were rather poor at the time. There was a significant setback in development of fish culture in rice fields in the 1960–1970s due to the overemphasis on crop production (Miao 2002).

In China, fish culture in rice fields entered a new development phase in 1980s. Total area of rice-fish culture (excluding the area used for fingerling production) exceeded 1 million ha by 1995, more than double that in 1982. The production of marketable fish from rice fields reached 273,000 tons with an average unit production of 265 kg/ha.

Since the late 1990s, another fundamental change has taken place in the development of rice-fish culture. Focus of fish culture in rice fields has shifted from production centered practices to more ecologically and socioeconomically focused patterns. Along with the relatively stable total culture area and production, various new culture systems and practices have developed and new species have been introduced to improve the ecological and socioeconomic benefits of production. Presently, fish culture in rice fields has developed into an important agro-aquaculture system that contributes not only to food security and rural livelihoods, but also to food safety and environmental integrity in China.

2.3 Evolution of Fish Culture in Rice Fields

Fish culture in rice fields is a traditional integrated fish farming practice. This traditional food production system has gone through a series of changes in the past three decades, which can be divided into a number of evolutionary phases. The farming practices have different features to address, and are also limited by the environmental conditions prevailing at the time.

2.3.1 Initial Development Stages

Fish culture in rice fields remained as a very primitive practice for a long period in China, though great importance was attached to developing fish culture in rice fields as early as late 1950s. Except for stocking, often of small size, there was almost no other input. Hardly any specific modification was made to the rice fields to meet any special requirements of culture of aquatic animals. Rice harvest was the main production in the system and the fish yield was very low. Due to the insignificant difference in price between fish and rice (fish price only doubled that of the price of rice only in early 1980s in China), the economic importance of fish in the production system was very limited due to its relatively small production and low price. Although the areas of rice cultivation increased, fish production from the systems did not keep pace. Fish production from rice fields was often impacted by the efforts to raise rice production, such as increased use of pesticides, chemical fertilizer, and introduction of new and high yielding rice varieties which were often not favorable to fish growth. Such a situation lasted until the mid-1980s.

2.3.2 Production Centered Rice-fish Culture

In mid-1980s, in order to meet the increasing demand of the people for aquatic products and offset the stagnant production from capture fisheries, both inland and marine capture fisheries, promotion of aquaculture, particularly inland aquaculture, became a very important development policy of the Chinese government. A lot of effort was made with the government support to develop rice-fish culture in China with a major emphasis to improve production efficiency. Standard rice field engineering for different types of rice-fish culture practices under different conditions (environmental, agronomic, social) and for different fish species was developed and disseminated to vast areas. The disseminated system is typically called a Ridge-Field-Trench system. Dimensions and proportion of different components of the system were standardized to meet the best needs of both rice and fish, and resulted in improvements in the yield of both rice and fish. It also made rice-fish culture widely applicable under different environmental conditions and greatly facilitated the expansion and transferring of the culture practices.

2.3.3 Rice-Fish Culture and Income Generation

Since the mid-1990s, increasing the income of rural populations and improving the livelihoods have become a top priority of the government development policies in China. A challenging task for governments at different levels in China (Central, Provincial, Municipal, and County) was to improve the economic returns from the land, so as to discourage leaving the land. Developing aquaculture, including rice-fish culture, was widely considered as an effective approach to achieve the above goal. Past practices of rice-fish culture with traditional species, mainly carps and tilapia, due to low farm gate price, could not meet such requirements fully. Many high valued finfish species and crustacean were, therefore, introduced to the systems and included swamp eel, pond loach, freshwater prawn, freshwater crab and freshwater crawfish, etc. Introduction of these high valued species significantly improved the economic return from rice-fish culture, though the unit production remains more or less similar to that in the past or even slightly decreased.

2.3.4 Holistic Ecosystem Approach for Environmental Integrity and Food Safety

Owing to the new trend of increasing concerns of the public about environmental integrity and food safety, many traditional food production systems including rice-fish culture, are confronted with new and more challenges. The means of producing high quality and safe rice and aquatic products through environmentally friendly approaches have become the focus of development of modern day rice-fish culture. Various research and experiments have been carried out to develop rice–fish eco-culture systems and models, which include design of rice–fish ecosystems for different target species, environmentally friendly farming management practices covering feeding and/or fertilization, pest and weed control, and water management. To ensure environmental integrity and food safety, many new technologies were introduced to the farming systems, such as the use of formulated bio-fertilizer and microagent for water manipulation. In many areas, traditional rice-fish culture has gradually transformed into green/organic food production systems. Certificated and labeled green and organic rice and aquatic products not only bring the farmers more income, but also contribute to holistic rural development.

2.4 Recent Developments

2.4.1 Trends of Change in Culture Area and Production

Since the mid-1980, fish culture in rice fields has achieved significant growth in China due to various reasons. The area of rice field with fish culture increased from 648,666 ha in 1985 to some 1.55 million ha in 2007 (MOA 2008; Fig. 2.1). The production of finfish and other aquatic animals from culture in rice fields increased from 81,699 tons in 1985 to 1.16 million tons in 2007.

The rapid growth of rice field fish culture (both area and production; Fig. 2.1) mainly took place during 1993–2002. This is largely due to the supporting policies of central and local governments to develop rice-fish culture as an important strategy to increase the income of rural farmers. However, expansion of area of rice–fish culture was less compared to other aquaculture systems. For example, fish pond

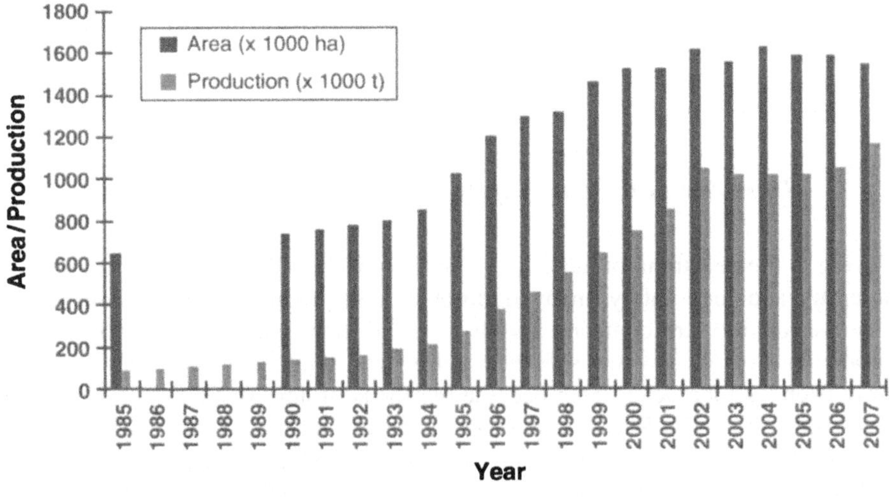

Fig. 2.1 Area of and fish production from fish culture in paddy fields in China (1985–2007)

area increased by 162.3% during 1983–2006 in China. In comparison, the area of rice-fish culture increased only by 139.2%. Although there has been a slight change in the area of rice–fish culture, the production has maintained a fairly steady trend of growth. This is a good indicator of improved management practices and culture efficiency. The production growth attained is not significantly higher in the recent past compared with the 1990s. This was largely due to the increase of area for culture of high valued species in rice–fish culture, such as freshwater prawn, crab and high valued fish (e.g. swamp eel), the unit production of which is usually much lower than the traditionally cultured carp species in rice fields. However, the economic return has been significantly improved as these species usually fetch 3–5 times higher price than carp species.

The total production of fish from rice–fish culture has maintained a steady growth trend for more than two decades, though the growth rate has varied in the different periods. In general, the production increase has been much faster compared to the area expansion of rice-fish culture. During 1985–2007, the total production increased by 13-fold. It is obvious that such an increase is achieved mainly through the improvement of production efficiency, which has resulted from both increased inputs and improved management practices. Present average unit fish production from rice-fish culture has reached about 780 kg/ha in China in contrast to 126 kg/ha in 1985.

2.4.2 Species Diversification

In the recent decades, diversification of species of aquatic animals cultured in rice fields can be considered as an important trend. Many indigenous and exotic species have been introduced to rice–fish farming systems. These species are usually more suitable for rice field environment, and are often of much higher market value. Currently, some 30 aquatic animal species are cultured in rice fields in different parts to China (Zeng and Wang 2006). Important species cultured in rice–fish system are given in Table 2.1.

2.4.3 Proliferation of Rice–fish Culture Across the Country

In addition to the general growth trends of rice–fish culture in the past three decades, there has been also extensive proliferation of the culture practices across China in the past two decades. In 1990, rice-fish culture was practiced in 20 provinces/MDUCG/Autonomous Regions in China. Only 11 provinces/MDUCG/Autonomous Regions had fish production from rice fields exceeding 1,000 tons. By 2007, rice-fish culture expanded to 25 provinces/MUCG/Autonomous Regions in China. There were 18 provinces/MUCG/Autonomous Regions that produced more than 1,000 tons of fish from rice fields. Fourteen provinces/MDUCG/Autonomous

Table 2.1 Important aquatic animal species cultured in paddy fields in China

Traditionally cultured species	Recently introduced species
Common carp, *Cypinus carpio*; Crucian carp, *Carasius auratus*; Grass carp, *Ctenopharyngodon idellus*; Silver carp, *Hypophthalmichthys molitrix*; Bighead carp, *Aristichthys nobilis*; White amur bream, *Parabramis pekinensis*; Blunt snout bream, *Megalobrama amblycephala*; Tilapia, *Oreochromis niloticus, O. niloticus* (♀) × *O. aureus* (♂); Catfish *Clarias* spp., *Pseudobagrus fulvidraco*	Swamp eel, *Monopterus albus*; Pond loach, *Misgurnus anguillicaudatus*; Oriental river prawn, *Macrobrachium nipponese*; Giant river prawn, *M. rosenbergii*; White-leg shrimp, *Penaeus vannamei*; Freshwater river crab, *Eriocheir sinensis*; Red swamp crawfish, *Procambarus clarkii*

Regions produced more than 10,000 tons of fish from rice fields, in contrast to only three in 1990.

There were many features associated with such proliferation of rice–fish culture. There has been a latitudinal expansion of the rice–fish culture practices, which, formerly, was mainly concentrated in low latitude areas, such as southwest, south central, and eastern parts of China. Rice–fish culture has been extended to the northeastern, northern, and northwestern parts of China. Topographic expansion of the culture practices from hilly and mountainous areas in the past to plain areas and outskirts of urban areas was also responsible for proliferation of rice–fish culture. Rice–fish culture has also expanded across economic boundaries. Rice–fish culture that was mainly practiced in economically undeveloped areas in China before has now become a very popular practice in many economically advanced areas. For instance, Jiangsu province is one of the most economically developed provinces in China. Rice–fish culture is now practiced in 90% of its over 60 counties.

2.4.4 Extended Scope and Well-organized Industry

Rice–fish culture was originally a very scattered small family-based operation. In the past two decades, rice–fish farming has gradually developed into a well organized business. With the general agriculture and aquaculture zoning, rice–fish culture often forms large scope industry in designated rice–fish farming areas with good infrastructure established by local governments. The production is still a family-based operation of farmers who obtain long term (30–50 years) leases from the government. The farmers are well organized through farmers associations or other forms. Marketing facilities are also specially established close to large rice–fish culture areas to facilitate marketing and obtaining reasonable prices.

The large scope operation of rice-fish culture in certain areas has a great advantage for local fisheries service agencies to effectively provide technical support to the farmers. It also facilitates the organization of production inputs and marketing of products.

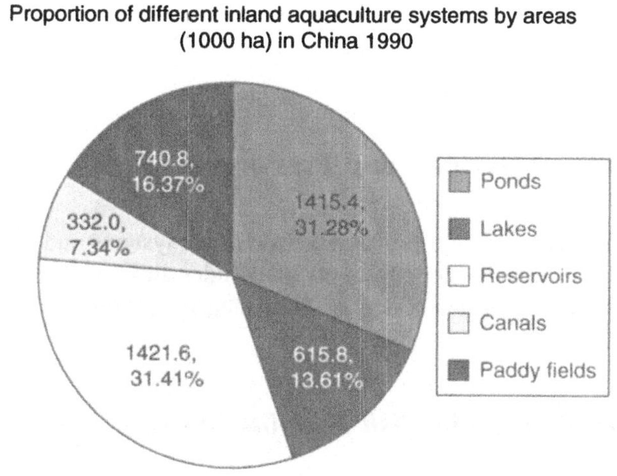

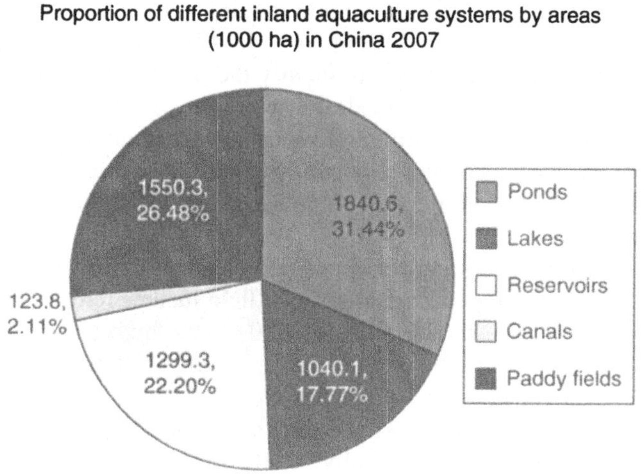

Fig. 2.2 The relative proportions of inland aquaculture systems in 1990 and 2007

2.5 Role of Rice-fish Culture in Inland Aquaculture

The rapid development of rice-fish culture in China has significantly upgraded the role of rice–fish culture production in inland aquaculture as well as the overall agriculture sector in China. This is well demonstrated by change in structure of inland aquaculture by systems between 1990 and 2007 (Fig. 2.2).

The share of rice field fish culture in total inland aquaculture areas increased from 16.4% in 1990 to 26.3% in 2007. It occupied the second important position in terms of area just after pond culture. Considering the reduced total area of rice fields, increasing area of rice–fish culture has upgraded its position in the general agriculture sector. Meanwhile, the contribution of rice-fish culture to inland aquaculture increased from 3.5% in 1990 to over 6% in 2007.

2.6 Important Systems and Practices

Along with the development processes, diversified systems and practices of rice–fish culture have been developed and adopted under different environmental conditions or for different species of aquatic animals across China.

2.6.1 Concurrent Rice-fish Farming Systems

Culture of fish concurrently with rice during more or less the same production period is a most typical rice–fish symbiotic farming system. Therefore, the potential ecological and economic benefits of rice–fish integration are assured. Accordingly, specialized earthworks to modify the rice fields to meet the requirements of both rice and fish are needed. It also requires more careful management in fertilization, pest/disease control, and water management. The suitable species for culture are those which do not require a long period to reach marketable size in grow-out culture. Concurrent rice–fish farming is very suitable for rice-crustacean culture practices.

Concurrent rice–fish farming requires good engineering earthwork of the rice field. Major earthworks for preparation of rice field include reinforcement of field ridge. The rice field ridge should be at least 10–20 cm higher than the maximum water level in the rice field. The rice field must be seepage free.

Peripheral and central crossing trenches are constructed in the rice field. The peripheral trench is usually 80–100 cm inside the rice field ridge to avoid escape of cultured animals. The trench is normally 50–100 cm wide and 60–80 cm deep. A central fish pit (5–6 m^2 and 1 m deep) connected to the trench is sometimes dug in the central area of the rice field in large rice plots (Fig. 2.3). The total area of fish trench and pit usually accounts for 10–15% of the rice field area in order to maintain more or less the same unit production of rice. In concurrent rice–fish culture system, the yield of fish usually ranges 300–900 kg/ha, and that of prawn and crab is in the range of 300–750 kg/ha. Production of crawfish is much higher.

The most important management aspects in concurrent systems include water management and use of pesticides and fertilizer. While meeting the requirement of rice at different stages, minimum water depth should be maintained for culture of animals in trenches and pits, 30–50 cm depending on species and weather conditions. Pesticides should not be used just before rain. It is important to avoid the pesticides getting into the water when spraying. Organic fertilizer is preferred if fertilization is needed. Use of ammonium fertilizers should be avoided as it often leads to high ammonia level in water, which is toxic to cultured animals.

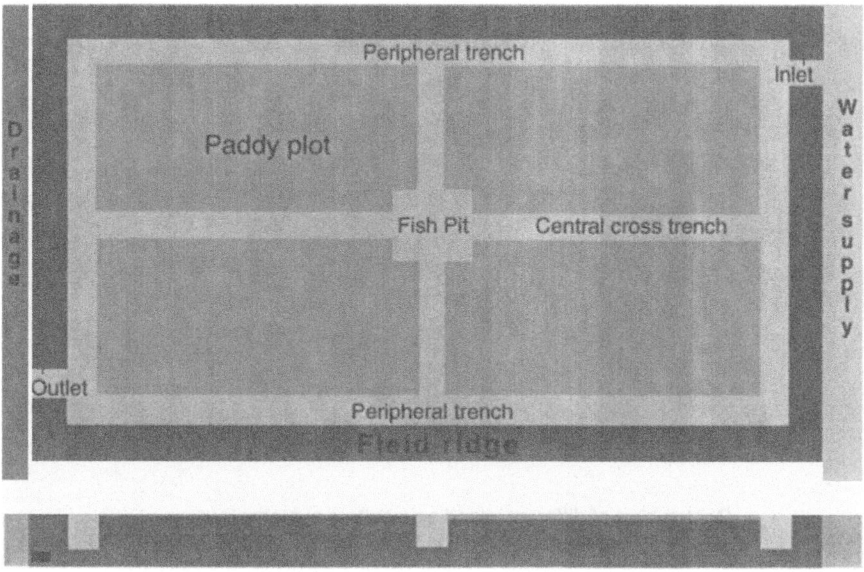

Fig. 2.3 Layout of ridge-plot-trench rice–fish system

2.6.2 Alternate Rice–fish Farming Systems

In some rice–fish culture systems, instead of concurrent rice–fish culture, rice plantation and fish culture are conducted alternately in the same field. Rice plantation and fish culture may be conducted alternately by crop or annually. Compared with concurrent rice–fish culture systems, rice and fish do not have close symbiotic relationship in alternate farming systems. However, alternative rice plantation and fish culture are beneficial to each other in many aspects. Alternative rice–fish culture is relatively easy and does not require extensive earthwork to modify the structure of the field. It is more suitable for fish species that require deep water and a longer culture period.

To conduct alternate rice–fish farming, the major requirements to rice field include reinforced rice field ridge, increased height of the field ridge (around 1 m), and a seepage free ridge.

The alternate rice–fish culture usually includes one crop of rice followed by one crop of fish within a year. The fish production usually ranges between 1,500 and 3,000 kg/ha or even higher depending on species and feeding. The stocking of fish is usually carried out immediately after the preparation of the field where the rice is harvested in October. The rice straw is buried into soil as fertilizer and feed for fish. The fish is usually harvested in June before transplantation of rice seedlings. In both concurrent and alternate rice–fish systems, water inlet and outlet are positioned diagonally and well fenced with screens of appropriate mesh-size depending on stocking size of the cultured animals.

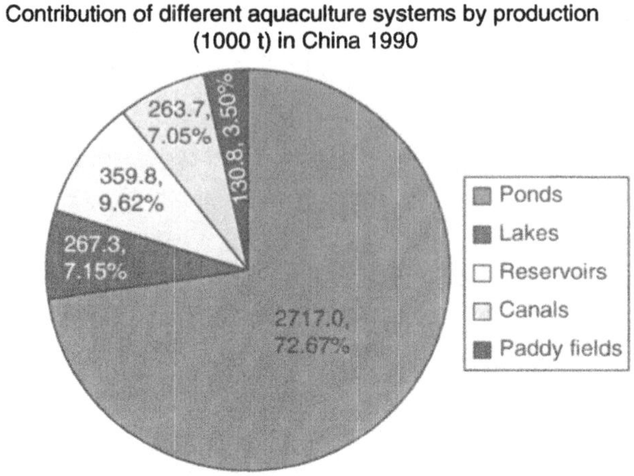

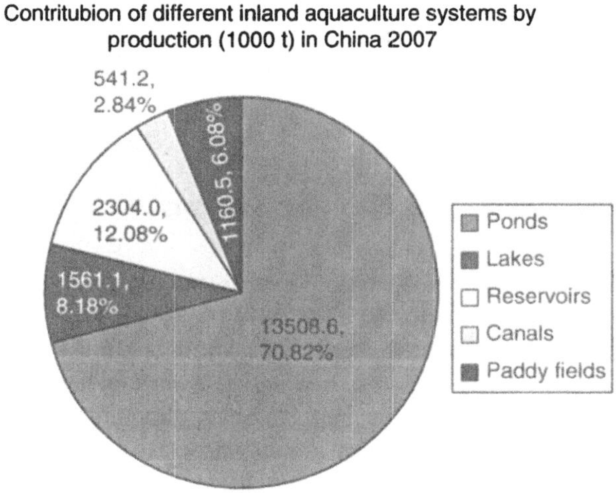

Fig. 2.4 Contribution of different inland aquaculture systems to overall production in 1990 and 2007

2.6.3 Rice–Finfish Farming Practice

In the ancient time, rice–fish culture started with the practice of concurrent rice–finfish integration. This type of practice had been continued for thousands of years with gradual modifications. Old practices of finfish species involved in rice–fish culture were mainly carp species, such as common carp, crucian carp, grass carp, silver carp, and bighead carp. These carp species are still important in present rice–fish farming systems. However, many new finfish species have been introduced to the practices (Plate 2.1), such as swamp eel, loach, catfish, and tilapia. These species are either having high market value or more suitable to rice environment.

Plate 2.1 Swamp eel-rice culture in Jingjiang city, China (Miao 2002)

Fish fingerlings of large size (8–15 cm body length) are usually stocked a week after transplanted rice seedling turns green. The total stocking density is usually 3,000–12,000 fingerlings/ha depending on supplementary feeding. Omnivorous species (common carp, crucian carp, and tilapia, for example) usually account for 60–80% of the total stocking, while herbivorous species (grass carp and blunt snout bream) account for 20–40% of the total stocking. Without supplementary feeding, 450–750 kg/ha of fish can be produced, and supplementary feeding is required for higher yields. Commonly used feeds include grass for herbivorous fish and fine rice bran, wheat, soybean cake, and rapeseed cake for both herbivorous and omnivorous species. Fish fingerlings are usually disinfected with saline water or bleaching powder solution before stocking. Liming is also commonly practiced during the preparation of the rice field at the rate of 450 kg/ha.

2.6.4 Rice–Crustacean Farming Practices

Crustaceans have been introduced to rice–fish farming just about a decade ago due to the relative short history of inland crustacean culture in China. Currently, important crustacean species involved in rice–fish culture include freshwater prawn (*M. nipponense* and *M. rosenbergii*), Chinese mitten handed crab (*E. sinensis*), red swamp crawfish (*P. clarkii*), and white-leg prawn (*P. vannamei*). Despite the short history, rice-crustacean culture has gradually become a mainstream practice in many parts of China, driven by high market value and better adaptability to the rice field environment. However, successful rice-crustacean culture requires different modifications to the rice field, particularly installation of shelter and more careful management practices, because crustacean are more sensitive to many chemicals and pesticides used in rice cultivation. Economic returns from rice-crustacean culture are much higher than rice–finfish culture. Normal net income usually ranges between $2,500 and $4,000 per hectare.

Red swamp crawfish-rice culture is a rapidly developing practice in the central part of China. Apart from the normal fish trenches and pits inside the rice field, vertical surrounding fence (30 cm deep into soil and 30 cm high above soil) established with bricks, rigid plastic sheets, or other materials is required for prevention of escape of crawfish. Aquatic plants are needed to be planted inside the trenches and pits to provide shelter for postmolting protection, which usually covers about 30% of the trench and pit area. Some 300–750 kg/ha of berried crawfish or 150,000–225,000 individuals/ha of juvenile crawfish (2–3 cm) are stocked. Supplementary feeding is required especially when stocking density is high. Crawfish is typically omnivorous. By-products of grain processing and oil extraction, animal feed such as snail, trash fish, earthworm, poultry entrails, and floating aquatic plants are all well accepted. The feeding rate is dependent on the feeding of crawfish, which should be monitored daily. It often ranges 3–5% of the body weight and is subject to adjustment based on the water quality and weather conditions. Feeding is much reduced when water quality is somewhat poor, or continuous rainy and cloudy weather is encountered. The stocking is usually carried out in November (for berried crawfish) or March (for juvenile). Harvesting is through both selective harvesting (trap set inside the trench and pit) and total harvesting by draining the water. The average yield is 6,000–7,500 kg/ha (Plate 2.2).

Plate 2.2 Rice-river crab culture in Nantong city, China (Miao 2002)

2.7 Contribution of Rice–fish Culture to Overall Aquaculture in China

As a unique production system producing two most important food items namely rice and fish, development of rice–fish culture has been making outstanding contribution to people's life and social development in China in different aspects.

2.8 Social and Economic Benefits

Social and economic benefits are the most important contribution of rice–fish culture development in China, though rice–fish culture is basically a food production activity. Social benefits of rice–fish culture development cover the following major aspects.

2.8.1 Rural Farmer Household Income

Chinese rice farmers are characterized by small scale family operations. The land farmed by one household is usually between 0.2 and 1 ha. The income level of the household is very much limited by the small land area with farming production of ordinary crops like rice, wheat, and maize. Farmers can hardly make a net profit of $100 from 1 t of grain, i.e., a family with 1 ha of land (the actual average farm is only 0.52 ha) may make a maximum of $1,500 net income with traditional crops annually when harvest is good. Therefore, it is impossible to significantly increase the income of rural farmers through growing only the traditional crops.

Rice–fish culture provides a much more effective approach to increasing the income of rural farmer households in China. The average unit production of 780 kg/ha fish from rice–fish culture can bring the farmer $1,500–4,000/ha income. The usual net income from good practice of rice–fish culture is $2,000–4,000/ha. The farmer's income can be increased by 2–4 times compared with sole crop farming. In general, income of some 2–3 million rural households has been significantly improved through rice–fish culture in China.

Jiangsu province in central eastern China sets a very good example of contributing rice–fish development to rural peoples' income. The province started a large project to promote rice–fish culture with high valued aquatic species in 1997. The total rice–fish culture area reached 136,615 ha in 1999. More than 80% of the area adopted rice–fish culture of high valued species. The gross net profit of rice–fish culture reached $2,912/ha on average. In Maxi village in Yancheng city, 146 ha of rice fields were utilized for rice-crab seed production, where the net profit reached over $11,000/ha in average (Wang 2000).

2.8.2 *Improvement of Women's Social Status*

Development of rice-fish culture can also contribute significantly to improvement of social status of women. Rice–fish culture production is a typical production system that provides equal opportunities for participation of both men and women. In practice, most of the daily management work of rice–fish culture is mainly carried out by women. Some housewives are involved in selling of the products. This significantly improves the participation of women in production activities and strengthens their role in household livelihood.

2.8.3 *Promoting Social Stability*

Rice–fish culture development has contributed to overall social stability in China. China is currently facing a huge population pressure. The surplus labor power and limited means of making good living in rural areas have stimulated a rapid flow of rural populations to urban areas. Urbanization can only be a gradual process in a large developing country like China because rapid movement of population from rural to urban areas may cause instability in both areas. Often the urban areas do not have the capacity to accommodate a rapid inflow of rural population. Development of rice–fish culture with sound economic returns provides rural farmers with an alternative means to significantly improve their living standards, and hence, induce them to stay in rural areas. It can ease the pressure on urban areas caused by the rapid migration of population from rural to urban, thereby contributing to overall social stability in China.

In addition, production activities involved in rice–fish culture enhance the good living style of the people in rural areas. It is a fact that crop farming activities on limited land require very limited labor inputs, so that surplus labor is available for other uses in rural agricultural communities. If this surplus labor is not effectively utilized, there is a risk of leading to unhealthy activities of rural people such as gambling which could create social problems. As rice–fish culture activities require more labor input for daily management, rural farmers can be better engaged in a rural economic activity. It, therefore, assists in diverting rural people from unhealthy activities to productive activities which can improve their living.

2.9 Contribution to Food Security and Safety

As an efficient production system of high quality food products, rice–fish culture is making substantial contribution to national food security and safety in China.

2.9.1 *An Effective Approach of Food Production with Minimum Natural Resource*

Rice–fish culture system is a typical way of producing more food through effective utilization of existing land resource. With good management practices, some 500–1,000 kg of fish products can be produced from one ha of rice field through concurrent farming system with very limited other material inputs.

Contribution of rice–fish culture to inland aquaculture production in 1990 and 2007 is presented in Fig. 2.4. The contribution of rice–fish culture to total inland aquaculture production in China has increased from 3.5% in 1990 to 6.08% in 2007. In 2007, rice–fish culture provided 1.16 million tons of high quality fish products to Chinese people with an increase of 8.87 folds since 1990. Considering the fast development of the overall inland aquaculture industry, this contribution is in fact very significant.

In addition to the increased fish production, well-managed rice–fish culture practices can also increase the productivity of rice. Usually, unit production of rice from rice–fish culture can be increased by 5–15% (Wu 2000). With the present area of 1.55 million ha of rice–fish culture, it is estimated that some additional 1 million tons of rice are produced every year.

2.9.2 *Effective System Producing Green and Organic Rice and Aquatic Products*

With the improvement of living standards of people, there is increasing demand for food of better quality, particularly safety. There is growing concern over the quality of rice and fish produced with extensive use of chemical fertilizers, pesticides, and aqua-drugs in intensive fish culture and rice cultivation. Market demand for green and organic foods is growing in China, which results in much higher price of green and organic food products than ordinary ones.

With the recent trends of development of ecosystem of rice–fish culture and good management practices of rice–fish culture, there is a tendency to minimize the use of chemical fertilizers, pesticides, herbicides, and various therapeutics in the production process. Ecosystem rice–fish culture has now become an important source of green and organic rice, and aquatic products. While meeting the demand of consumers, it also brings the farmers' additional income through much higher market value. This has become an effective incentive for rice–fish farmers to shift to environmentally friendly ecosystem production practices.

2.9.3 Contribution to Environmental Integrity

Environmental sustainability is attracting increasing concern worldwide in the new millennium. Development of rice–fish culture in China has contributed significantly to the environmental integrity.

2.9.4 Minimizing Environment Impacts Through Least Use of Chemicals and Drugs

Ordinary rice cultivation and intensive fish culture often involve use of a range of chemicals and drugs for the control of diseases, pests/parasites, and weeds. The use of large quantities of chemicals and drugs not only affect the quality of the product, but also cause adverse impacts on the environment with residues added to water and soil. Rice–fish culture creates a symbiotic system where fish play an effective role in the control of pests and diseases of rice in the system. Experiments and production practices have shown that incidence of rice diseases/pests is significantly reduced in rice–fish culture systems (Table 2.2). The application of pesticides can be lowered by 50% of that of modern high-input rice production; sometimes, no pesticide application is required (Lu and Li 2006). Fish and other aquatic animals also help in weed control of the system. For example, when 1 kg of grass carp is produced in a rice field, about 80 kg of weed is consumed by fish (Zeng and Wang 2006). As a result, the use of pesticides and herbicides is significantly reduced in rice–fish culture system. Herbicide is only used before the transplantation of rice and weeding throughout the rice growing period is not necessary.

On the other hand, incidence of fish diseases is also considerably reduced compared with other aquaculture systems, particularly in pond culture. This is largely due the feature of the rice system.

Hence, the environment impacts involved in rice and fish production are minimal if not zero, while producing high quality rice and fishery products. Hardly any other food production system can be compared with rice–fish system in this regard.

Table 2.2 The effect of bio-control of rice pests and disease in rice–fish ecosystems

Rice treatment	Rice plant hopper	*Naranga aenescens*	Moore *parnara guttata*	Rice leaf roller	Rice leaf hopper	Sheath rot (%)
Monoculture	237	534	20	11	839	37.7
Rice–fish	75	107	10	6	227	4.9

Unit: head/patch^{-1} (number of pest in one testing patch) (Lu and Li 2006)

2.9.5 *Maximum Utilization of Production Inputs and Discharge of Nutrients to the Environment*

Sole rice plantation and intensive fish culture usually depend on use of large quantity of fertilizers and feeds. The conversion efficiency in either rice system or fish culture system is often far from satisfactory and results in high production costs. Moreover, it increases the nutrient loading to natural water bodies when discharged, which might result in algal blooms and seriously hamper the normal functions of aquatic systems. The problem is even more serious than industrial pollution now in China because control of nutrient discharge from rice cultivation and fish culture practices is much more difficult than industrial pollution.

In the symbiotic system of rice–fish culture, on the other hand, living organisms produced in the rice field through fertilization can be utilized by fish. This also significantly reduces the need for supplementary feeding for fish. The fecal matter of fish also serves as rich organic fertilizer for rice, which reduces the inorganic fertilizer needs for rice. With a fish production of 375 kg/ha in rice fields, the feces from the fish are equivalent to a fertilization of 93.8 kg/ha of ammonium sulfate and 33.8 kg/ha of calcium superphosphate (Zeng and Wang 2006). The nitrogen-fixation role in the system also increases the organic matter content, total nitrogen, and total phosphorus in the soil by 15.6–38.5% (Lu and Li 2006).

Ultimately, whatever fertilizers or feeds used in the system are efficiently utilized and converted into food production, and nutrients discharge to natural environment is minimized. rice-fish culture system also reduces the emission of CH_4 by nearly 30% compared to traditional rice farming (Lu and Li 2006).

Chinese government is implementing more stringent control of waste discharge from various production systems, including aquaculture. Water quality standards for effluent discharge are being established and will be implemented soon. In future, rice-fish culture will, therefore, have a better advantage over intensive pond fish culture in this regard.

2.9.6 *Maximizing Water and Land Resource Use*

Although China has a high volume of freshwater resources, per capita availability of inland water is very low due to high population. As such, water conservation is an important national policy in China towards sustainable development.

Both rice cultivation and fish culture are water dependent. Practice of rice-fish culture maximizes the utilization of water resources because it saves water required for fish production.

Similar to inland water resources, China is also a country with very low per capita availability of arable land. As such, wise-use of arable land resources is one of the key national development policies in the country. Currently, arable land is no longer permitted to be converted into fish ponds. Development of rice-fish culture is, therefore, one of the few options for further expansion of aquaculture in China. Through the rice–fish integration, particularly the concurrent farming system, utilization of land resources is maximized.

2.10 Key Factors Contributing to the Success of Rice-fish Culture

In the past three decades, rice-fish culture has witnessed fast development and created significant socioeconomic benefits in China. Many factors have contributed to this development. The followings are among the most important ones.

2.10.1 Cultural and Socioeconomic Basis

Cultural and socioeconomic factors associated with rice cultivation and fish farming are important aspects to be considered in rice-fish culture. China has a long history of both rice cultivation and fish culture. Lands producing both rice and fish were considered as a paradise on the earth in the ancient times. Even in the present day, both rice and fish remain as important food items to ordinary Chinese people. China has a vast area under rice cultivation which is suitable for rice-fish culture, and there are abundant low cost manpower resources in rural areas. The market demand for fish products is increasing with the improvement of people's living standards. All these form solid cultural and socioeconomic basis for development of rice field fish culture in China.

2.10.2 Changing Social and Economic Environment

Development of production sectors is generally related to the social and economic environmental changes, which often raise new requirements/demands for changes in production sectors. During the past three decades, China has passed through a number of important development stages. The social needs in each development stage have positively stimulated development of rice–fish culture in China.

China started its economic reforms in early 1980s. The government's priority at that time was to boost the production of major food products, particularly animal food products, to meet the increasing demand. Fish products are among the most important animal foods of traditional significance in China and fish supply from capture fisheries encountered a stagnant production commencing in the early 1980s. Rice–fish culture production system was promoted by the government as an important means to increase fish production, especially for increasing the supply of fish to rural areas. This stimulated a rapid development of rice–fish culture from early 1980s to early 1990s.

Since mid 1990s, China has been successful in addressing issues of food supply for the people. Fisheries products generally meet the demand of the consumers. The Chinese government then faced another challenging task of increasing income and improving the livelihood of its huge rural population effectively. The advantages of rice–fish culture over the traditional crop farming have made it one of the priority choices of government at different levels for increasing rural farmer income. As improvement of economic efficiency of rice–fish culture was also a major focus, many new rice–fish models with high valued species were developed and disseminated widely.

In the new millennium, environment integrity in development and food safety have become pressing issues attracting increasing concerns worldwide. With regard to environmental integrity and food safety, rice–fish culture system has an additional advantage for further development. Refinement of technology and management practices of rice–fish culture has been promoted to meet the requirement of producing green and organic food through a more holistic development approach and ecosystem production system. Hence, rice-fish culture system in China was successful in grasping the opportunity and responding positively to such demand by restructuring and refining the production system essentially driven by emerging external needs.

2.10.3 Changing of People's Perception

Transformation of simple, traditional rice–fish culture practice into modern rice–fish culture system often brings about difficulties to traditional rice–fish farmers at initial stages due to their traditional ways of thinking. It was, therefore, important to change their perceptions toward modern technology in order to enable the rural rice farmers to understand the principles and benefits of new production systems and practices. Such changes can only be done through good demonstration of the advantages of the new and advanced systems/practices and other knowledge and information sharing mechanisms. This process must be facilitated by the public service systems supported by the government.

2.10.4 Favorable Government Policy Steering and Public Support

For most developing countries, any social, economic, and industrial changes need strong government policy steering, especially for a large country with diversified social and environmental conditions. It is the responsibility of the government to develop favorable policies to support and encourage the sector to make needed changes and to meet the new/emerging requirements of a changing society. Such policies should be developed and implemented at different levels of government and public institutions.

The success of rice–fish culture development is largely because the Chinese government developed appropriate policies and strategies to promote the development. Most importantly, such policies have been effectively implemented at the grass-root level.

In developing countries, rural development is often associated with large number of small scale producers, often family-owned and operated. Strong public support is, therefore, needed to facilitate the development and upgradation of a food production sector or systems. Successful development of rice–fish culture in China has greatly benefited from such support.

The public support, which has facilitated the successful development of rice–fish culture, includes the followings. Many government-supported research and extension institutions have endeavored in development and demonstration of new rice–fish culture models and related management practices, which can lead to better economic efficiency and environment benefits. Capacity building for promoting good rice–fish practices has been supported by technical training of extension workers at different levels and farmers in most provinces where rice–fish culture has achieved significant development. Government-supported large scale technology dissemination programs have played a major role in the development process. Such programs usually involve not only technical support, but often provided financial support, particularly for improvement of infrastructure and production facilities. Public support in facilitating the marketing of the rice–fish farmers has also contributed to the success significantly. It is vitally important to enable the farmers easily market their fish products from rice as it is often much more difficult than dealing with crops and it is much needed for rice–fish production developed to a commercial scale. Public marketing facilities and other related infrastructure have been established as one of the measures to support the rice-farmer in China. Establishing the physical infrastructure and facilities for fish marketing with government funding support not only benefited the rice–fish farmer, but also generated revenue to local government for other public welfare organizations.

2.11 Way Forward

2.11.1 Potential for Further Development

2.11.1.1 Further Expansion of Culture

China is the largest rice producer in the world and there are some 25 million ha rice fields in the country. About 10 million ha of rice fields are considered suitable for fish culture (Meng and Wu 2002). Currently, only 15% of the area is used for rice-fish culture, and therefore, there is a great potential for further expansion of rice-fish culture.

(a) Improvement in Production Efficiency

There has been a significant progress in production efficiency improvement in the past decades. Nevertheless, the range in the unit production level is still relatively high across the country presently. With the average national fish yield of 781 kg/ha from rice fields in 2007, on average, the highest unit production reached 2,466 kg/ha in Shanghai Municipality directly under the central government (MDUCG). In contrast, the lowest average unit production from rice-fish culture was 160 kg/ha in Chongqing MDUCG. Such large differences in unit production in rice-fish culture are largely caused by the difference in technical and managerial practices, though it may also be influenced by environmental differences to some extent. Therefore, there is tremendous potential to raise the production efficiency through the dissemination of modern culture techniques and advanced management practices.

2.11.1.2 Scaling Up the Production of Green and Organic Food

Currently, production of green and organic rice and fish products through environmentally friendly rice–fish culture practice is still in the early stage in China. Practice of environmentally friendly rice–fish culture requires more technical inputs and careful management practices. As the educational level of majority of rice–fish farmer is rather low, greater efforts should be made to conduct awareness programs for the farmers about the practice of environmentally friendly rice–fish culture and the potential benefit of such kinds of practices. Further expansion of environmentally friendly rice–fish culture will bring the rice–fish farmers greater economic benefits, and will contribute more to the food safety and environmental sustainability.

(a) Challenges to Scaling Up and the Coping Strategies

Although potential of further development of rice–fish culture is great in China, full realization of the development potential is dependent on how the following challenges are addressed effectively.

(b) Environment Constraints

China has enjoyed fast economic development in the past three decades, which has greatly improved the living standards of Chinese people. At the same time, the rapid economic development has been achieved at an environmental cost to a certain extent, such as resource overexploitation and environmental degradation. Water scarcity and quality issues are quite serious in some areas of the country. Although major actions have been taken by the government at different levels in China to address these environmental issues, significant improvement to environmental quality is a rather slow process. Good quality water supply can be a key constraint to further development of rice–fish culture in certain areas of the country as ecosystem of rice–fish culture is more dependent on the supply of good quality water. Concerted efforts are, therefore, needed to restore and improve environmental quality.

(c) Improvement of the Public Service for the Small Farmers

Rice–fish farming involves large number of small scale farmers. Traditional rice farmers often have limited knowledge about rice–fish culture. Strong technical support services are, therefore, required for further expansion of rice–fish culture and upgrading of present practices. In the restructuring of local government and public service institutions, less number of technical staff supported by local government is maintained at grass-root level resulting in high workloads of such personnel. As such, the extension mechanisms are still rather inefficient in upgrading and scaling up rice–fish culture production.

On the other hand, economic return to farmers from rice–fish production is highly dependent on market changes. Small scale rice–fish farmers have poor access to reliable timely market information, which often results in difficulty to market the products at reasonable market prices and profit margins. In a market-oriented economic system, predicting market trends and guiding rice–fish farmers based on these predictions are other challenges to the public service institutions.

It is, therefore, time to establish public market study and information distribution systems for small-scale farmers. Promotion of farmers' associations at different levels is apparently an effective approach to assist individual farmers in addressing marketing problems. Government facilitation and sharing of successful experiences across the regions and sectors are very important.

(d) Leading Rice–fish Farmers Towards Ecosystem Practices

Environmentally friendly ecosystem rice–fish culture practices have not been widely adopted among rice–fish farmers due to the limited knowledge and management skills of the majority of rice–fish farmers. Some farmers are more influenced by short-term benefits. Empowerment of farmers with required knowledge and skills for shifting from traditional rice–fish practices to more ecosystem rice–fish culture is a major challenge.

It is very important to establish a conducive climate for socially and environmentally holistic development approach through public advocacy. In addition to demonstration of good practices of rice–fish culture and sharing of successful lessons among farmers and across regions, market forces should be used to drive small scale rice–fish and rice farmers towards better practices of ecosystem rice–fish culture practices. The production of organic rice and fish should be encouraged in rice–fish systems, which would add economic value to the system and also raise farmer income (Lu and Li 2006). A scheme of certification and eco-labeling for green and organic fish products and rice, and development of the market demands will be helpful to lead rice–fish farmers towards adopting ecosystem practices.

2.12 Regional Perspectives

Asia is the most important area of rice and fish production. Both rice and fish products play an extremely important role in food security and livelihoods of rural populations in the region as a whole. The rice-fish culture systems, therefore, have a great potential in rural development in the region. With some 140 million ha rice fields (nearly 90% of the world total), the potential of rice-fish culture development is immense.

Rice-fish culture has been practiced in many countries in Asia. Nile tilapia seed is produced in rice fields in Bangladesh (Barman and Little 2006). Performance of common carp, *C. carpio* L. and Nile tilapia, *O. niloticus* (L.) in integrated rice–fish culture was also assessed in Bangladesh (Frei et al. 2007) and found that carp/tilapia mixed culture with supplementary feeding would maximize the output from rice–fish culture. In India, 0.23 million ha of rice fields out of a total of 20 million ha are considered to be suitable for the adoption of rice–fish integration systems (Mohanty and Mishra 2003). In northeast Thailand, the major species stocked in rice fields are the silver barb (*Puntius gonionotus* = *Barbonymus gonionotus*), common carp, and Nile tilapia (Little et al. 1996). In addition, rice-fish culture is also practiced in other Asian countries, such as Cambodia, Myanmar, Indonesia, and the Philippines. However, rice-fish culture has not received sufficient attention of governments and the public in most Asian countries. In general, the production scope is relatively small and the production efficiency is rather low and is often limited by the lack of constant supply of fish seed and impacted by the use of pesticides.

The favorable climatic conditions in many Asian countries for rice–fish culture also indicate a high potential for rice–fish culture development. Establishing a regular supply of fish seed and introducing ridge-plot-trench systems might enable the rice–fish farmers in the region to improve production efficiency. Efforts should be made by the public sector and NGOs to demonstrate the advanced rice-fish culture systems, models, and environmentally friendly management practices to the rice farmers. New species of aquatic animals of high value having better adaptability to rice cultivation conditions should be introduced to improve the economic returns to farmers. In conclusion, with the abundant natural and human resources facilitated with appropriate strategies, rice-fish culture will have a bright future to make greater contribution to rural development in the region.

References

Barman, K.B. and D. Little. 2006. Nile tilapia (*Oreochromis niloticus*) seed production in irrigated rice-fields in Northwest Bangladesh – an approach appropriate for poorer farmers? *Aquaculture* 261: 72–79.

Ministry of Agriculture. 2008. *China fisheries statistic yearbooks 1985–2008*. Beijing, China: Bureau of Fisheries, Ministry of Agriculture.

dela Cruz, C.R., C. Lightfoot, B.A. Costa-Pierce, V.R. Carangal, and M.P. Bimbao. eds. 1992. *Rice-fish research and development in Asia*. ICLARM Conference Proceedings, vol. 24, 457. Manila: ICLARM.

El Gamel, A.R. 1997. Egyptian aquaculture: Status and development requirements, with special emphasis on tilapias and their potential in aquaculture. In *Tilapia aquaculture*, ed. K. Fritzsimmons, 141–153. Ithaca, NY: ISTA, NRAES.

FAO. 2008. *FAOSTAT-agriculture production 2007*. Rome, Italy: FAO.

Fernando, C.H. 1993. Rice field ecology and fish culture: An overview. *Hydrobiologia* 259: 91–103.

Fernando, C.H. 1996. Ecology of rice fields and its bearing on fisheries and fish culture. In *Perspectives in Asian fisheries*, ed. S.S. De Silva, 217–237. Manila: Asian Fisheries Society.

Fernando, C.H. and M. Halwart. 2000. Possibilities for the integration of fish farming into irrigation systems. *Fisheries Management and Ecology* 7: 45–54.

Fernando, C.H. and M. Halwart. 2001. Fish farming in irrigation systems: Sri Lanka and global view. *Sri Lanka Journal of Aquatic Sciences* 6: 1–74.

Fernando, C.H., F. Golthenboth, and J. Margraf. 2005. *Aquatic Ecology of Rice Fields*. Kitchener, Ontario: Volumes Publishing. 472.

Frei, M., M.A. Razzak, and M.M. Hossain. 2007. Performance of common carp, *Cyprinus carpio L.* and Nile tilapia, *Oreochromis niloticus* (L.) in integrated rice–fish culture in Bangladesh. *Aquaculture* 262: 250–259.

Halwart, M. 1998. Trends in rice-fish farming. *FAN* 18: 3–11.

Li, K. 1988. Rice-fish culture in China: A review. *Aquaculture* 71: 173–186.

Little, D.C., P. Surintaraseree, and N.I. Taylor. 1996. Review-fish culture in rainfed rice fields of northeast Thailand. *Aquaculture* 140: 295–321.

Lu, J. and X. Li. 2006. Review of rice–fish-farming systems in China – one of the Globally Important Ingenious Agricultural Heritage Systems (GIAHS). *Aquaculture* 260: 106–113.

MacKay, K.T. (ed). 1995. *Rice-fish culture in China*. Ottawa: IDRC. 264.

Meng, X. and W. Wu. 2002. New development of rice–fish culture in China. *China Fisheries Economics* 1: 17–19.

Miao, W. 2002. Analysis on status of rice-fish culture development in China and recommendations for further development. In *Collection of investigation papers on fisheries development policy in China*, 83–87. Beijing: Bureau of Fisheries, Ministry of Agriculture.

Mohanty, R.K. and A. Mishra. 2003. Successful rice–fish integrated farming adopting rainwater conservation measures. *Fishing Chimes* 22(10–11): 97–99.

National Statistic Bureau. 2008. *China statistic yearbook 2007*. Beijing: China Statistic Press.

Pullin, R.S.V., and Z.H. Shehadeh. 1980. *Integrated farming systems*. ICLARM Conference Proceedings, vol. 4. Manila: ICLARM.

Shehadeh, Z.H. and L. Feidi. 1996. Aquaculture and resource limitations in Egypt. *FAN* 14: 3–7.

Szito, A. and R.W. Jones. 1997. Carp culture in rice fields in Hungary. In *Integrated fish farming*, ed. J.A. Mathias, A.T. Charles, and B. Hu. New York: CRC Press.

Wang, Y. 2000. Full recognition of importance of role of rice–fish culture and push forward its development. *Scientific Fish Farming* 9: 8–9.

Zeng, Z. and S. Wang. 2006. Analysis to development history of rice–fish culture and the driving factors. *Journal of Nanjing Agricultural University* 6(3): 79–83.

Chapter 3
Shrimp Farmers in India: Empowering Small-Scale Farmers through a Cluster-Based Approach

N.R. Umesh, A.B. Chandra Mohan, G. Ravibabu, P.A. Padiyar, M.J. Phillips, C.V. Mohan, and B. Vishnu Bhat

Abstract The great bulk of shrimp farming in India, as in most of Asia, as well as that of aquaculture in general in the region, is based on small scale farming activities, and in this regard, is no exception to other primary sector activities. The work on the development of better management practices (BMPs) on the shrimp culture sector commenced with the recognition of the need to place the sector on a firmer footing, while combating the problems of frequent disease occurrences, and to ensure its long term sustainability. The process commenced with the organization of small scale farmers into groups – clusters and/or aquaclubs – grouping farmers in a given area, drawing on common resources such as a common water supply channel, and inducing the farmers to act collectively rather than individually to the betterment and benefits of all. Such clusters and/or aquaclubs were later transformed into Societies with a legal standing, with the establishment of the National Center for Sustainable Aquaculture in 2007, with a purview to monitor society functioning and dissemination of technical know-how to other areas. The outcomes include improved shrimp yields, less impact on the environment, improved product quality, and better relations among players in the market chain. In short, the society formation and/or organization of small scale farmers into groups facilitated the adoption and implementation of the BMPs providing benefits to the farmers, environment, and society. As a result of the cluster approach, and hence, adoption of BMPs, shrimp production has increased from 4 tons in 2002 to 870 tons in 2006. Implementation of simple, science-based farm level plans (e.g., BMPs) and adoption of cluster farming through the participatory concept reduced disease risks in cluster farms significantly, for example, in the demonstration farms, it was reduced from 82% in 2003 to 17% in 2006. Economically, for every Rs. 1,000 (US$ 25) invested by a farmer, around Rs. 520 (US$ 13) was earned as net profit in 2006. This was a substantial increase compared to the Rs. 250 (US$ 6) profit made by nondemonstration farmers during the same period.

N.R. Umesh (✉), A.B.C. Mohan, G. Ravibabu, P.A. Padiyar, M.J. Phillips, C.V. Mohan, and B.V. Bhat
National Centre for Sustainable Aquaculture, 69-17-8, SBI Officers Colony, Rajendra Nagar, Kakinada-533003, AP, India
e-mail: nacsa.hq@gmail.com

S.S. De Silva and F.B. Davy (eds.), *Success Stories in Asian Aquaculture*,

The organization of small scale shrimp farmers in India has (a) empowered small-scale farmers, (b) increased stakeholder interaction and involvement within the clusters, and (c) is an ideal model for small scale farmers to meet market requirements, a model accepted by many Asian countries and being increasingly adopted. The model is of a self propagating nature, and most of all, has contributed to the sustainability of shrimp farming in India, and indeed, in the region. Needless to say, this example drawn from India best exemplifies how small scale shrimp farmers could remain economically viable and are able to comply to the increasing demands of sophisticated markets, and most of all, achieve sustainability.

Abbreviations

CAA	Coastal Aquaculture Authority
MPEDA	The Marine Products Export Development Authority
NACA	Network of Aquaculture Centres in Asia Pacific
NaCSA	National Centre for Sustainable Aquaculture
NATURLAND	Association for Organic Agriculture was founded in 1982 with its headquarters in Graefelfing, near Munich, Germany
SIPPO	Swiss Import Promotion Programme

3.1 Introduction[1]

The Network of Aquaculture of Centers in Asia Pacific (NACA), in collaboration with the Marine Products Export Development Authority (MPEDA), Government of India, conceived and implemented a project for "Shrimp disease control and

[1] Definitions used in this presentation are as follows:

- *Better management practices (BMP)*: Management practices aimed at improving the quantity, safety, and quality of products taking into consideration animal health and welfare, food safety, and environmental and socio-economic sustainability. BMPs are management practices, and implementation is generally voluntary; they are *not* a standard for certification. The term "better" is preferred rather than "best" because aquaculture practices are dynamic and continuously improving (today's "best" is tomorrow's "norm").
- *Cluster*: Cluster is a group of farmers whose shrimp ponds are situated in a specified geographical locality; commonly all ponds are dependent on the same water source.
- *Aquaclub*: Aquaclub is a term used in India to describe an informal group of farmers cooperating with each other on various aspects of management in the cluster.
- *Society*: Society is a term used in India to describe a formal (legal) registered group of (20–75) aqua farmers in a farming locality. Societies are setup according to a model established by the Indian government, and they are legally registered by the Ministry of Revenue and subject to annual audits by government officials to verify accounts and ensure a democratic and transparent management.

coastal management" to address disease and environmental problems in the shrimp industry in India, and ensure that small shrimp farmers of India meet high standards for biosecurity, food safety, and environmental protection. The project aimed to address capacity building in shrimp health and quality management at the grass-roots level by organizing small scale farmers into aquaculture clusters.

The shrimp industry is a key sector in India's economy because of its significant contributions to export earnings and gainful employment. Given its vast natural water resources, India has tremendous potential to excel in aquaculture. But in reality, the country's shrimp exports had stagnated since the late 90s. Problems began a few years earlier with the outbreak of white spot disease (WSS). Later on, the judgment by the highest court of India on shrimp aquaculture also impacted its advancement. In response, to address the rising concerns about the sustainability of the sector, in the year 2000, the MPEDA with the technical assistance of NACA initiated the aforementioned project. The project started in 2001 with a large-scale epidemiological study aimed at identifying the risk factors for key shrimp diseases and developing and disseminating better management practices (BMPs) to minimize farm-level risk factors for disease outbreaks and address more broadly shrimp farming sustainability.

The project has since been institutionalized to organize small shrimp farmers and build capacity at grassroots level in India, and provides a strong basis for future progress, as well as an example for other countries in addressing some of the special problems and concerns facing small scale aquaculture farmers.

3.2 The Shrimp Industry in India

The development of more commercial hatcheries, coupled with credit facilities from commercial banks and technical and financial assistance programs from MPEDA, led to a phenomenal increase in the area under shrimp farming between 1990 and 1994. The tiger prawn, *Penaeus monodon* indigenous to the region, was the main species cultured. In the states of Andhra Pradesh and Tamil Nadu, a large number of commercial integrated shrimp farming units with foreign collaboration also emerged adopting "scientific" culture systems with facilities for production of shrimp seed, shrimp feed, and processing. But this trend did not continue for long as the large scale corporate shrimp farms failed to make profits, and consequently, shrimp farming became more or less a small farmer activity. Presently, coastal aquaculture in India is synonymous with shrimp aquaculture and mainly carried out by small scale farmers. Shrimp farming in India involves three categories of farmers/entrepreneurs, i.e., small and medium farmers, middle-level entrepreneurs, and big entrepreneurs. The farming of shrimp is largely dependent on small holdings of less than 2 ha; these farms account for 90% of the total area utilized for shrimp culture, 7% of farms are between 2 and 5 ha and the remainder has an area of greater than 5 ha (Jayaraman 1998; MPEDA 2008). The small scale farmers were unorganized and most of the farmers did not have access to technological innovations

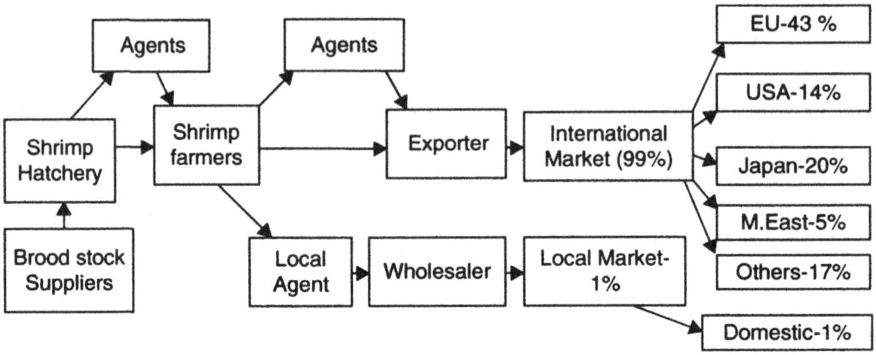

Fig. 3.1 The value chain for farmed shrimp (source MPEDA, 2008)

and scientific applications. They contribute around 80% to the total shrimp production, but were poorly served. Small-scale farmers are innovative and productive, but because of poor organization, lack of skills, inadequate information, and knowledge base, they are vulnerable to the numerous risks and hazards that impact their livelihoods, farm productivity, and competitiveness.

Shrimp farms are operated using both leased out government/private lands and landowner-operated shrimp farms. A credit system functioned throughout the sector, operated and controlled primarily by intermediaries. Intermediaries also acted as input suppliers and providers of credit at each stage in the supply chain, and were also involved in buying back the harvested shrimp. On average, farmers end up paying a whopping 30% interest on the loans from the intermediaries that affect the profitability of their operations.

At present, about 283 shrimp hatcheries operate in the country providing a total production capacity of 1,254 billion PL/year. Farmed shrimp production increased from 40,000 tons in 1991–1992 to 143,170 tons from a 140,000 ha farming area, and another 42,820 tons of scampi (the giant freshwater prawn, *Macrobrachium rosenbergii*) from a 4,300 ha area during 2006–2007, generating about Rs. 40,790 million in export sales equivalent to US $ 0.8 billion (MPEDA 2008). The value chains associated with the shrimp farming sector is shown in Fig. 3.1.

3.3 Background to MPEDA-NACA Project

In November 1994, the WSS, which is caused by a systemic ectodermal and mesodermal Baculovirus, led to extensive damage to shrimp culture in India (as elsewhere in Asia), and continued its devastation in 1995 and early 1996 throughout the west and east coasts of India. Both *P. monodon* and *P. indicus* in extensive, improved extensive, and semi-intensive farms, irrespective of the source of water for culture, were affected by the disease and the total loss was estimated to be about Indian rupees 6,000 million (Vijayakumaran 1998). With improved biosecurity measures, the severity of the WSS syndrome has since decreased in India and shrimp farming was limping back to normalcy until a court order issued by the Supreme Court of India in December 1996 suspended coastal aquaculture within 500 m from the high tide level in coastal regulation zone. This decree also led to the establishment of an Authority to regulate the development of coastal farms and farming practices. Subsequently, the Government of India enacted the Coastal Aquaculture Authority (CAA) Act, 2005, enabling the establishment of the CAA for enforcing proper regulatory measures for coastal aquaculture in a more sustainable and eco-friendly manner. The Act encompasses all forms of aquaculture in saline or brackish waters in the coastal areas, but excludes freshwater aquaculture.

While analyzing the issues concerned with small farmers, it was amply clear that the majority of the shrimp farmers neither had the skills to adopt scientific norms, nor access to useful technical information essential for shrimp farming. The awareness levels of farmers were inadequate and neither the Government nor the farmers were geared to meet the challenges that were posed by issues, such as pollution, viral diseases, and traceability and food safety concerns. The availability of technical personnel in the fisheries departments in the respective states to support the vital extension functions at the grassroots level were inadequate, resulting in poor transfer of technology, lack of coordination with other departments, and poor research linkages. With the conventional top-down approaches showing limited success in extension services, there was a need to promote the bottom-up participatory approach with effective coordination and convergence at the appropriate levels.

In order to address the rising concerns about disease and sustainability of the sector, in the year 2000, the MPEDA with the technical assistance of NACA initiated a project on "Shrimp disease control and coastal management."

3.3.1 Key Steps Adopted in the Project (also see Fig. 3.2)

- *2000*: A baseline study of the major diseases affecting the shrimp aquaculture operations.
- *2001*: Longitudinal epidemiological study in 365 ponds in Andhra Pradesh to identify major risk factors associated with WSS and low productivity in *P. monodon* culture ponds.
- *2002*: Development of farm level contextualized BMPs to address the identified risk factors. Pilot testing of BMPs in selected farms.
- *2003:* Development and testing of the concept of cluster farming for effective BMP adoption among farmers in a cluster.
- *2004*: Expansion of BMP promotion to a large number of clusters. Extension of some of the BMPs to downstream activities such as hatcheries.
- *2005*: Review and refinement of BMPs and production of BMP extension leaflets for each stage of the culture operation.

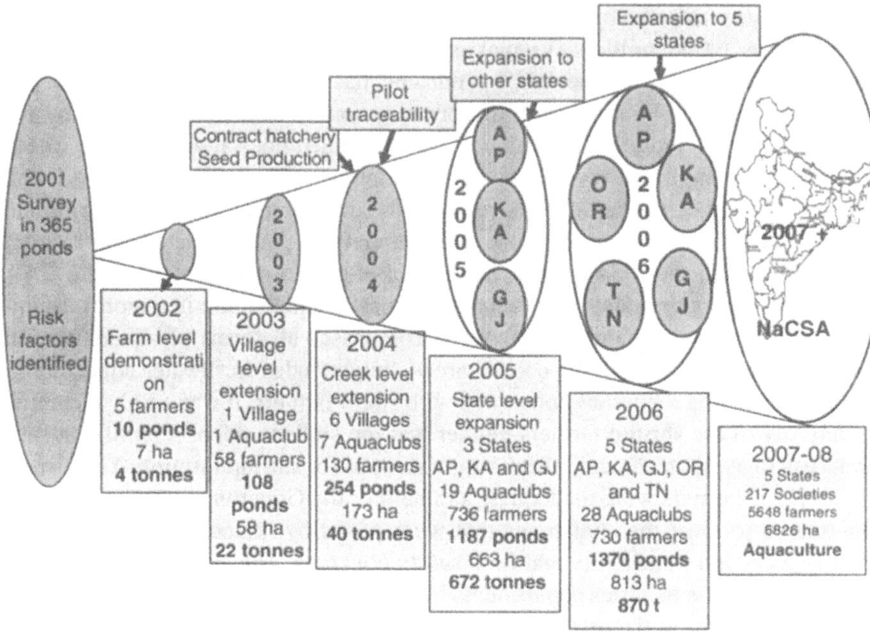

Fig. 3.2 MPEDA/NACA Project development since 2000. The figure indicates the adoption of the strategies by different states and increasing number of farmer groups (*A* – Andhra Pradesh, *G* – Gujarat, *K* – Karnataka, *O* – Orissa, *T* – Tamil Nadu)

- *2006*: Expansion of the BMP program to clusters in five different states. Conceptualization of an institutional framework for sustaining the cluster approach as aquaculture societies for sustainable aquaculture.
- *2007*: Establishment and inauguration of National Center for Sustainable Aquaculture (NaCSA) to carry forward the MPEDA/NACA project activities.

3.3.2 Risk Factor Study (2000–2001)

Following the base line survey and study of the major diseases that were affecting shrimp farming operations and also with some preliminary consultations with farming communities and other stakeholder groups, a detailed longitudinal epidemiological study was conducted on shrimp disease problems during 2001. The study involved 365 ponds in West Godavari and Nellore districts of Andhra Pradesh state, and used an epidemiological approach to better understand the key risk factors contributing to shrimp disease outbreaks and low pond production, with an emphasis on the economically impacting "WSS." The findings from the risk factor study and management practices to address the problems have been published into a "Shrimp Health Management Extension Manual" (MPEDA/NACA 2003).

3.3.3 Development of BMPs Based on International Principles for Responsible Shrimp Farming

The outcomes provided a better understanding of the risk factors for WSS outbreaks, and those causing unusually low pond productivity. Toward the end of 2001, results were discussed widely with farmers and other agencies in Andhra Pradesh, and some consensus was reached on the study findings and their practical application to improve performance of shrimp farming systems of Andhra Pradesh. Risk factors significantly associated with disease outbreaks and low pond productivity were then used to develop locally relevant management strategies BMPs to reduce the identified risks.

The MPEDA/NACA project in India has also provided a good example of translating the International Principles (FAO/NACA/UNEP/WB/WWF 2006) for responsible shrimp farming into BMPs adapted to local farming conditions, and ensuring their easy implementation by relevant stakeholders. The findings from these projects provide evidence of the advantages of small farmers being organized (into aquaclubs or societies), sharing resources, helping each other, and adopting BMPs. The outcomes include improved shrimp yields, less impact on the environment, improved product quality, and better relations among players in the market chain. In short, the implementation of the BMPs has provided benefits to the farmers, environment, and society (Mohan et al. 2008).

In order to enhance BMP uptake and promote adoption in different coastal states of India, BMP brochures on ten key thematic areas were developed in English and translated into the five regional languages of the respective states. For each of the thematic areas, the illustrated brochures describe the field procedures in 15 simple steps in the local language. The BMP dissemination was principally through weekly farmer meetings, society coordinators, and regular pond visits by the NaCSA staff.

The key BMPs developed and implemented in the project are:

1. *Good pond/water preparation*: The soil should be checked for the presence of black layer and should be removed from the pond. Water should be screened at the water inlet point to avoid entry of virus carrying fish and crustaceans, which may be predator or competitor for shrimp. Water depth of at least 80 cm should be maintained in the pond.
2. *Good quality seed selection*: Seed is best purchased through contract hatchery seed procurement system where all the group farmers purchase quality seed for whole group.
3. *Water quality management*: Basic water quality parameters like dissolved oxygen, pH, and alkalinity must be maintained at optimum level. Water exchanges only when felt necessary and during critical periods.
4. *Feed management*: Efficient use of feed. Demand feeding using check trays and feeding across the pond using boat/floating device. FCR must be kept below 1:1.5.
5. *Pond bottom monitoring*: The pond bottom soil should be monitored on weekly basis, especially at the feeding area or trench for black soil, benthic algae, and bad smell, and corrective actions should be taken.

6. *Health monitoring/biosecurity*: No draining or abandoning of disease affected stocks. Farmer groups are encouraged to discuss common actions that can be taken during disease outbreaks to avoid spreading of disease from one farm to another.
7. *Food safety*: No use of any harmful/banned chemicals like pesticides and antibiotics.
8. *Better harvest and post-harvest practices*: Quick harvesting, chill killing of harvested shrimp, and quick transport to processing plant.
9. *Record maintenance/traceability*: Maintenance of hatchery/pond management record book by hatcheries and farms to identify problems in the tank/pond environment, and to rectify these problems at the earliest during the production cycle. This is also required for traceability purpose.
10. *Environmental awareness*: Improved environmental awareness about mangroves, pollution, and waste management among farmers.

3.4 Implementation of BMPs and Evolution of the Group Approach

3.4.1 Pilot Testing of BMPs at Farm Level (2002)

In 2002, demonstrations were conducted in five selected private farms, involving ten ponds, in three villages in West Godavari and Nellore districts of Andhra Pradesh on the east coast of India. NACA and MPEDA provided technical assistance to demonstration farmers for on-farm testing of BMPs, and supported monitoring and evaluation to understand benefits and constraints. The demonstrations were also used more widely to disseminate information on risk management strategies to farmers. Although the adoption of BMPs did not completely eliminate shrimp disease problems, the outcomes as judged by participating farmers and the MPEDA/NACA study team were very promising. Adoption of pond level risk management practices led to improvements in both profits and productivity. In demonstration farms, returns shifted from a loss in 80% of ponds in 2001 to a profit in 80% of ponds in 2002. During district level workshops in November 2002, with over 470 farmer participants from Nellore and Bhimavaram, farmers responded positively to the findings, and requested urgent support for more demonstration activities and initiatives to extend the concept of BMPs to the wider farming community.

3.4.2 Promotion of BMPs at Group Level (2003)

In 2003, MPEDA and NACA responded positively to farmer requests and supported an extension of the project for further demonstrations in one farmer group in a village of West Godavari District of Andhra Pradesh. The objective was to promote adoption of BMPs across a wider number of farmers to create a visible and quantifiable impact on the village community through organization of a "self-help group" (aquaclub) to collectively address common shrimp health and farm management problems using a participatory approach (collective planning, decision making, and implement crop activities).

The core NACA/MPEDA team lived in the village during the early part of 2003, promoting adoption of BMPs, supporting farmers to establish the Aquaclub, facilitated weekly farmer meetings, and organized "service provider – farmer" contacts and exchange of information, thereby trying to build up mutual trust among these parties. At the same time, the team established a monitoring program and at the end of the 2003 crop, evaluated with farmers the outcomes of the village demonstration to better understand the benefits and constraints to adopting better health management practices.

3.5 Promotion of BMPs at the Cluster Level and Expansion of the Program (2004–2006)

Following the success of BMP promotion at the group level, the program moved one step higher and promoted BMP adoption among clusters along a creek (their shared water source). In 2004, 130 farmers with 254 ponds were assisted to organize into seven aquaclubs/clusters in Andhra Pradesh and BMPs were promoted at the level of clusters (MPEDA-NACA 2005). The results of all of the above are best illustrated schematically when the number of farmers adopting BMPs increased almost exponentially within half a decade, and the individual profits obtained increased significantly (Fig. 3.2). Most of all, the farmer clusters have become sustainable and the livelihoods of small scale farmers that were in jeopardy previously have been ensured by scaling up and institutionalization of the program.

The MPEDA-NACA project was very successful, and the enthusiasm of the farmers involved in the project motivated MPEDA to expand this mechanism for capacity building in shrimp health and quality management at the grassroots level by organizing small scale farmers into Aquaclubs and clusters. MPEDA invested in a separate Technical Service Agency called the NaCSA under the administrative control of MPEDA. NaCSA has become operational since April 2007, and is functioning as an outreach organization of MPEDA primarily to cater to the extension needs of small scale aqua farmers. The primary objective of NaCSA is to support development of sustainable aquaculture in India through efforts primarily aimed at empowering the marginalized and poorest of the poor in the aquaculture sector, besides disseminating technologies and information on better practices, sustainable and judicious utilization of the resources, use of science in day to day activities, marketing of the produce, etc., for the benefit of the shrimp farmers and sustainability of the shrimp sector in the country (Fig. 3.3).

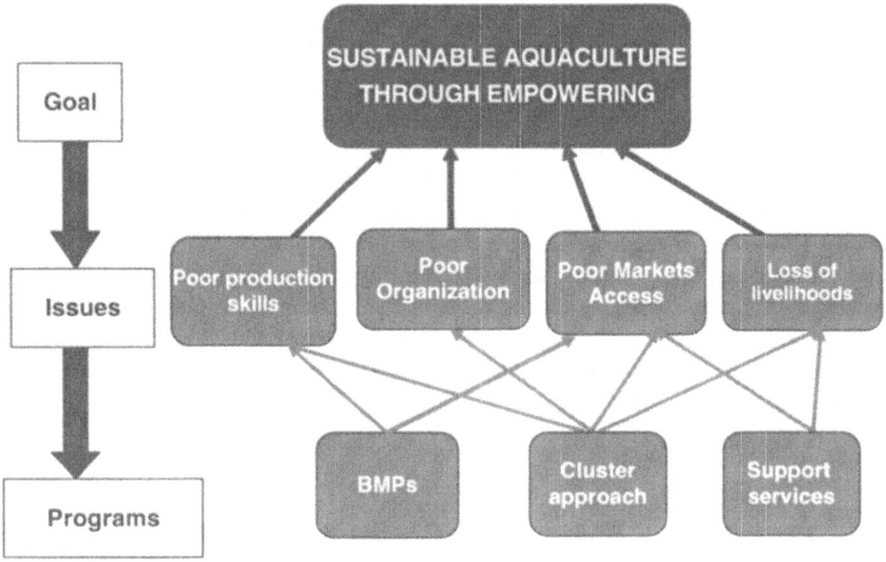

Fig. 3.3 Schematic representation of NaCSA mission and programs

NaCSA continues to consolidate and expand the BMP implementation work through the cluster concept initiated under the MPEDA-NACA program across the country through a network of aquaculture societies.

In the preparation and institutionalization phase, the NaCSA facilitated the formulation of a work plan by involving multistakeholders consultation in September 2007. The meeting was attended by over 35 delegates representing Indian Council of Agriculture Research, state fisheries departments, farmers, farmer organizations, hatcheries, processors, feed companies, professional organizations (SAP), PCR laboratories, and NACA. From MPEDA, Chairman, Director, and senior officers were present.

Now farmer groups are referred to as societies, which were called as aquaclubs earlier. Societies are setup according to a model established by the Indian government, and are registered by the Ministry of Revenue and subject to annual audits by MPEDA to verify accounts and ensure a democratic and transparent management. In a society, all the members register their farms with CAA and obtain a license. Each society consists of 20–75 farmers with strict conditions for membership and elected board members, and is a legal entity. The societies are also eligible for availing financial assistance from MPEDA and other agencies for various activities related to farming.

Intensive work in pilot clusters by project team members has led to the gaining of expertise. Moving this knowledge beyond pilot clusters required scaling up through human resources recruitment. NaCSA, accordingly, has recruited additional 20 field staff, training them and disseminating the project findings to more farmers. Also developed were organizational structure and work mechanisms and increased public awareness through various media and direct farmer meetings.

Andhra Pradesh is taking the lead in society organization as most of the BMP work was done in this state over the last 5 years. At present, NaCSA is working with more than 150 aquaculture societies, and expected to promote BMPs in 500 societies within the next couple of years.

3.5.1 Why the Approach is Considered a Success

A number of factors are thought to have contributed to the success of the approach adopted in the present exercise. Foremost among these are:

(a) Empowering small-scale farmers
(b) Ideal model for small scale farmers to meet market requirements
(c) Contributing to sustainability of shrimp farming
(d) Increased stakeholder interaction and involvement within the clusters
(e) Acceptance of the model by other NACA member countries
(f) Self propagating nature of the model

3.5.1.1 Empowering Small-Scale Farmers

Organized farmer groups (societies) are one of the key mechanisms for supporting farmer empowerment. They have the potential for cooperative action, which can change the position of the farmer in relation to the opportunity structures, and thus, influence the business environment of the farming community. Moreover, small-scale farmers can, through organization, gain the advantages of economy of scale in accessing services and markets, which are otherwise limited to large commercial farmers. Farmer groups also improve information exchange and sharing among group members. The small scale shrimp farmer groups of India are in a better position today to gain these benefits compared to the situation when they were unorganized.

3.5.1.2 Society Management

Each society has its own guidelines and implements them. The societies are audited every year by MPEDA for the implementation of guidelines and BMPs, so far 52 societies have been audited and found to be implementing BMPs. Societies which fail to implement them would lose their society registration. Some features of the registered societies are as follows:

- A society consists of 20–75 farmers and has a clear organization with strict conditions for membership, and elected board members.
- Membership to a society is purely on voluntary basis.
- In a society, all the members register their farms with the CAA.
- The members contribute an admission fee of Rs. 1,000 (US$ 25), and in addition, members have to pay 0.5% of their revenue to the society corpus fund.

- The optimum stocking density for each cluster is decided and abided by farmers/ members.
- Seed purchase is through contract hatcheries 45 days prior to stocking.
- All the cluster farmers stock at the same time (within 2 weeks period).
- Agreement by all society farmers for not using any antibiotics and no/reduced chemical use.

Each of the farmer societies has one coordinator selected among its members or from the community by society farmers with a prescribed minimum education level. The society coordinator is trained in society management, BMPs, and extension techniques by NaCSA. The coordinator will be responsible for implementing BMPs in societies, and act as link between society farmers and NaCSA. Each of the NaCSA field managers will coordinate and manage the activities of ten such societies. MPEDA's society scheme provides partial financial assistance for farmers to employ a society coordinator for the first 2 years. The working mechanisms of NaCSA are schematically represented in Fig. 3.4.

To facilitate farmer involvement, ensure commitment, and inculcate confidence, the previous MPEDA-NACA and NaCSA field staff stay closer to farmer societies for the entire cropping season. Key farmers from other villages where MPEDA/ NACA, NaCSA had worked previously are invited to new villages to share their experiences. Wherever possible, field visits are arranged for farmers to other villages for first hand information exchange among farmers. A contract hatchery seed procurement system is followed in registered societies. Farmers' field days are organized at the end of successful crop cycles in societies to spread the message of success to more farmers.

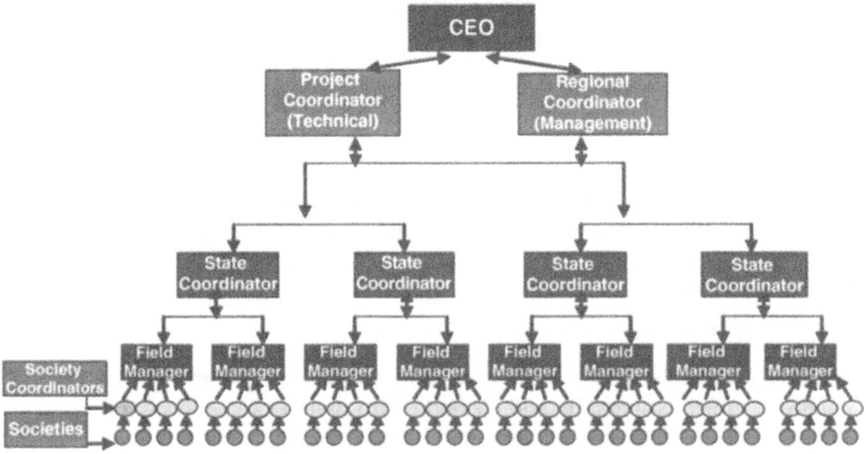

Fig. 3.4 NaCSA farmer society management plan

3.5.1.3 Capacity Building

NaCSA has taken up an extensive awareness program through village level meetings, which are key in educating farmers about market requirements and promoting the benefits of implementing BMPs through organized societies. During 2007–2008, a total of 251 village level meetings involving more than 5,000 farmers were conducted. The concept of BMPs and its implementation through society formation and the market requirement are explained in detail to all the farmers in a given area. Small farmers who do not have any other alternative other than shrimp farming are the ones who are keen to adopt BMPs through organized societies. In the process, we are seeing an emergence of farmer leaders in each society who are willing to work for the benefit of the group, many of them are voluntarily sparing their time and resources. In a village, there will be some aquaculture producers who are society members, while others are not members and prefer to keep conventional production and marketing practices. However, this trend is changing with the societies achieving better crop successes with reduced production costs, and having an opportunity to market their produce at better prices. All these have induced more and more farmers from a village to join either an existing society or form a new one.

There are 30 well-trained technical staff with NaCSA, besides 42 society coordinators trained by NaCSA. Professional Fisheries Graduates are also being trained in societies in various aspects of aquaculture, social, and environmental issues. NaCSA is planning to expand the same in future involving all the fisheries colleges in India.

3.5.2 Contributing to Sustainability of Shrimp Farming

For the aquaclubs/societies to be successful, they have to be economically and environmentally sustainable besides being socially responsible.

3.5.2.1 Economic Sustainability

Economic sustainability in aquaclubs/societies is achieved through

- Reduced disease risks
- Reduced cost of production and increased profits
- Improved service provisions

3.5.2.2 Reduced Disease Risks

The project has made significant progress, increasing from five farmers who adopted the cluster farm approach in 2002 to 730 farmers (813 ha) in 28 aquaclubs in five states (Andhra Pradesh, Karnataka, Orissa, Gujarat and Tamil Nadu) in 2006. The production of BMP shrimp through the program has increased from 4 tons in 2002 to 870 tons in 2006. Implementation of simple, science-based farm level plans (e.g. BMPs) and adoption of cluster farming through the participatory concept reduced disease risks in cluster farms significantly. The prevalence of shrimp disease in the demonstration farms was reduced from 82% in 2003 to 17% in 2006 (Fig. 3.5), while in nondemonstration ponds, the reduction in disease prevalence was limited during the same period.

In 2006, the program was run in five coastal states, namely, Andhra Pradesh, Karnataka, Gujarat, Tamil Nadu, and Orissa. BMPs were promoted in 28 clusters (aquaclubs) comprising 730 farmers with 1,370 ponds. Participation of farmers in BMP implementation during 2006 included:

- 20 aquaclubs in Andhra Pradesh
- 3 aquaclubs in Tamil Nadu
- 2 aquaclubs in Orissa
- 2 aquaclubs in Gujarat
- 2 aquaclubs in Karnataka

Compared to surrounding nondemonstration ponds, the crop highlights included:

- 30% increase in production
- 8% increase in size of shrimp
- 30% improvement in survival
- 31% reduction in disease prevalence
- Disease risks are reduced mainly through good quality seed

Good quality disease free seed is purchased through a contract hatchery system in which society farmers place bulk orders, 45–60 days in advance of the planned stocking date. Through a consultative process, initially facilitated by NaCSA, a mutual agreement is reached between selected hatcheries and societies. These agreements include compliance on the use of BMPs in hatcheries and other terms and conditions for production and procurement of quality seed (Padiyar 2005).

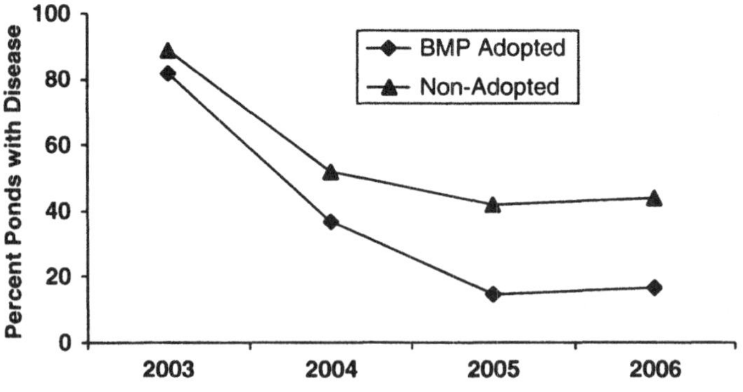

Fig. 3.5 Comparison of the prevalence (%) of disease in BMP and non-BMP shrimp ponds

- Same time stocking with optimum density
- Health management stress on prevention, rather than treatment
- Cooperation among farmers

3.5.2.3 Reduced Cost of Production and Increased Profit

Economic analysis of 2006 data clearly demonstrated that farmers adopting BMPs have higher profitability, lower cost of production, and are able to produce quality and traceable shrimp without using any banned chemicals. In the demonstration ponds for every Rs. 1,000 (US$ 25) invested by a farmer, around Rs. 520 (US$ 13) was earned as profit in 2006 (Fig. 3.6). This was a substantial increase compared to the Rs. 250 (US$ 6) profit made by non-demonstration farmers during the same period.

Prior to forming aquaclubs/societies, farmers never perceived in improving (e.g., de-silting the canal) the common water intake, but now with collective effort and pooling of resources, this has not only been made possible, the members appreciate the importance of regular de-silting to improve production. Now some farmers are progressing toward getting electricity for the entire cluster, which otherwise would have been beyond the means of individual farmers. Recognizing farmers' interest, Government is also coming forward to help the groups for development of infrastructure facilities.

Key factors responsible for improved profit in BMP implemented farms are:

- Efficient use of resources (feed)
- Reduced chemical use
- Sharing of expenses (deepening of canals, seed testing, transport of inputs, laboratory analysis, electricity, etc.)

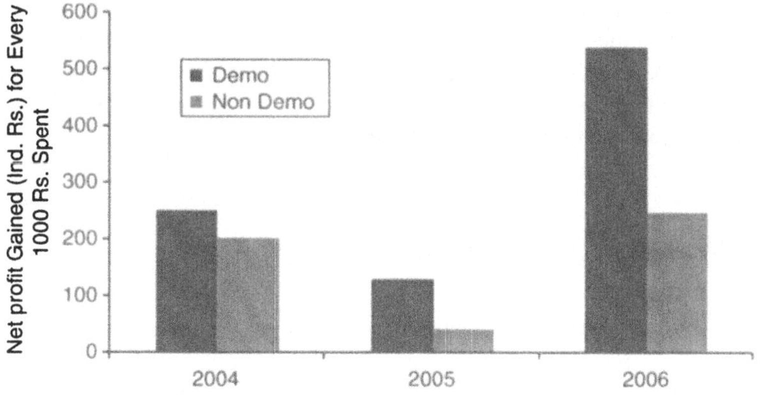

Fig. 3.6 Profit made by BMP (demo) and non-BMP farmers for every thousand rupees (US$ 25) investment

3.5.2.4 Facilitating Favorable Policy Decisions

The initiative has helped to create change towards policies that are more favorable to the small scale shrimp farmer. The Ministry of Commerce and Industry has approved a scheme for implementation through MPEDA on registration of Aquaculture Societies for adoption of code of practices for sustainable shrimp farming, with a total outlay of Rs. 25,000,000 (US$ 625,000) during the 10th and 11th plan period.

In the state of Andhra Pradesh, as soon as 100 societies were registered, all the society farmers gave a representation to the Chief Minister of the state requesting to intervene in helping small scale shrimp farmers with favorable policy changes.

3.6 Environmental Sustainability

The program also led to reduction in other aquaculture-related risks. The environmental risks were also reduced by the decrease in pollution resulting from efficient use of resources (energy and feed), reduced use of chemicals, antibiotics, and limited discharge of sediment and water exchange. In addition, the following initiatives have also been taken which will have positive environmental impacts.

3.6.1 Organic Project

The eco-friendly low density aquaculture practices of society farmers encouraged MPEDA to implement the first Indian Organic Aquaculture Project, which is a collaborative project of MPEDA and the Swiss Import Promotion Program in two aquaculture societies of Andhra Pradesh. MPEDA is providing financial assistance for the farmers to meet part of certification and expenses toward organic feed. The products will be certified by Naturland (www.naturland.com) against Naturland organic standards, providing market access with premium price. NaCSA is coordinating the implementation of the project in Andhra Pradesh.

3.6.2 Revival of Abandoned Ponds

Until mid-1990s, shrimp farmers earned good returns, but investments on technologies leading to good management practices were generally ignored. Consequently, shrimp farming in many areas failed to withstand the impact of viral outbreak in this period. As the situation failed to improve, a large number of farmers abandoned shrimp farming. In Krishna District of Andhra Pradesh where 1/3 of the farming area developed was abandoned, NaCSA took up demonstration with the objective of reviving the livelihoods of those farmers, in particular small scale farmers. Three societies consisting of 63 farmers (84 ponds, 67 ha) agreed to follow the BMPs starting with sourcing disease free seed through the contract hatchery system. After 110 days of successful culture, farmers started harvesting and none of the ponds of the three societies were affected by disease. More than 50% of non-society ponds were affected with WSS in this area during the last summer season, while 95% of the BMP adopted ponds made good profits during the period. Seeing the success of these societies, more and more neighboring farmers from abandoned areas are coming forward to organize themselves into societies. Positive impact of this success will be seen in coming crops as new societies implement BMPs and more societies will be organized in Krishna District. This could pave the way for full-scale revival of most of the abandoned ponds in Andhra Pradesh and an example for other parts of the (Table 3.1).

3.6.3 *Social Responsibility*

The social impacts of group farming include reduction in risks to livelihoods and improved awareness of biosecurity and environment integrity among cluster farmers. The key indicators of increased social responsibility among cluster farmers are:

- Regular information sharing among farmers during weekly meetings.
- Cooperation in selecting/testing and buying seed through contract hatchery seed production systems.
- All farmers in a cluster stocking at the same period, thereby avoiding continuous stocking and harvest.
- Reduced contamination when there is a disease outbreak due to information sharing among cluster farmers followed up with immediate remedial actions.
- Increased cooperation in sharing common facilities-deepening inlets, drains, etc.

Table 3.1 Details of farmer Societie organized to date (2008)

State	Organized societies	No. farmers	Area (ha)
Andhra Pradesh	121	2,492	2,697
Tamil Nadu	22	665	1,270
Karnataka	05	120	197
Orissa	03	49	55
Gujarat	02	53	38
Total	153	3,379	4,257

3.7 Opportunities to Comply (for Small Scale Farmers) with Market Requirements

Over the years, the approach to quality management has assumed greater significance and importance in the seafood sector worldwide, both in production and supply chains. New trends are emerging in production and marketing, such as traceability, eco-labeling, and certification. For farmers and producers in developing countries, supplying goods for national and international markets can present a life-changing opportunity. Retailer demand is there especially for products with ethical and green credentials. The difficulty lies in meeting those retailer needs and identifying the right products where developing country producers often lack the skills to deal with the high demands of the export markets and access to capital and business expertise. These factors collectively present a formidable barrier in entry to sophisticated markets. At the other end of the supply chain, retailers often lack the ability to be able to reliably source quality products that are required.

Farmer societies provide a good opportunity to link up with retailers following the existing controls that ensure that basic requirements of markets, including social and environmental responsibility and food safety, are in place.

- Traceability back to shrimp farms and hatcheries through proper record keeping and use of GIS maps.
- Legally registered farms – in India all shrimp farmers are registered by the CAA.
- Societies follow BMPs to control the hygiene and safety of shrimp produced by registered farmers.
- Society produced shrimp are safe – no use of antibiotics at any stage in society farms.
- Existing societies organized following the model of the Indian government and controlled by government officers, which have a high degree of compliance with fair-trade requirements in terms of democracy and transparency.
- Members of societies are familiar with export requirements.

3.8 Adoption of the Model by Other NACA Member Countries

Although BMPs are often simple farm level plans to prepare for and respond to disease, their systematic adoption by farming communities and countries to manage shrimp health problems and achieve widespread sustainable shrimp production has a relatively recent history. The MPEDA/NACA project has the distinction of being the first program moving in this direction in the region. Since then, this approach towards sustainability has been adopted by several countries, and it is expected to spread to many other countries in the Asian region and be suitably modified and adopted for other cultured commodities such as marine fin fish.

The following are the examples of positive uptake in countries in the region.

Vietnam has used the "International Principles" to adapt legislation and develop its national program toward better management of shrimp farming. In addition to supporting the development of the International Principles for Responsible Shrimp Farming, projects were initiated to translate the principles into practices, which targeted better production, product quality, and environmental and socio-economic sustainability. Among the government's initiatives to promote a more sustainable development of the sector was the project that supported coastal aquaculture, which demonstrated the private and social benefits of adopting BMPs. In 2003, NACA and the Ministry of Fisheries (MOFI; currently designated as Ministry of Agriculture and Rural Development – MARD) with the support of the DANIDA-funded Fisheries Sector Program Support (FSPS) began implementing a project to support the promotion of responsible shrimp farming at all levels and for all links in the production chain.

BMPs were developed for broodstock traders, hatcheries, seed traders, and farmers. Focus was given to the development of simple and practical BMPs, which addressed the needs of less resourced small-scale farmers. Ten sets of extension material were developed and disseminated in close collaboration with the MARD. The tangible outcomes include the following:

- Implementation of BMP for hatcheries was supported in six hatcheries and resulted in seed production up to 1.5 times higher and a price per unit seed of about 30–40% higher than non-BMP seed.
- BMP implementation was also supported in seven pilot farming communities (655 direct beneficiaries). Implementation led to a remarkably lower risk of mortality, higher production, and higher probability of making a profit.
- Farming communes that introduced seed testing increased their chances of making a profit of over 7 times.
- BMP application led to average yields that were sometimes more than 4 times higher than in farms where BMPs had not been adopted.
- The project BMPs were also incorporated into the draft standards for the production of organic seed.
- The project also strengthened the institutions involved with seed health management by conducting training courses and supporting the development of national and provincial-level legal documents to improve the process of seed screening and certification.

Benefits of BMP application in Vietnam were visible from the early stages of its implementation. Farmers complying even with only two recommended practices – testing of seed for WSS Virus and removal of sludge before stocking, reduced the risk of crop failure from 61.0 to 47.8%.

In Indonesia, BMP experiences from India were used in the rehabilitation of the shrimp farming sector in the Province of Aceh, following the 2004 tsunami. A practical BMP manual was prepared during 2006 based on the International Principles for Responsible Shrimp Farming, and the manual has been widely promoted and used by various agencies involved in assistance to rehabilitation of livelihoods in Aceh. The results from practical implementation are also promising, with similar outcomes of reduced disease risks and improved productivity in traditional shrimp farms compared to farmers not adopting better practices.

3.9 Increased Stakeholder Interactions and Involvement

The message from this project is guided by the importance of government support and collaboration of various institutions and partners to translate Principles into Practice. By coming up with the project ideas and committing resources to their implementation, the government of India provided opportunities for other local, national, regional, and international institutions, organizations, and agencies to take part in these projects. In India, MPEDA, State Department of Fisheries, ICAR and its relevant institutions particularly the Central Institute of Brackish water Aquaculture (CIBA), All India Shrimp Hatchery Association, Farmers' Associations, Seafood Exporters Association of India, academic institutions like the College of Fisheries, Mangalore, ACIAR, and FAO all had various roles to play.

Farmers being the primary producers, there is a need to link them to all other stakeholders in the industry both backwards and forward. Farmers are being linked to hatcheries, input suppliers, processors, scientists, Research Institutes, Government institutes, banks, and others. Bank loans for working capital, which are not available now for most of the small scale farmers, are likely to be made available once the societies are linked up with the market. The operational links are depicted schematically in Fig. 3.7. MPEDA is extending financial assistance in the form of the society scheme to kick-start the formation of the clubs and implement the BMPs. There is better flow of valuable information from field to research institutes. During

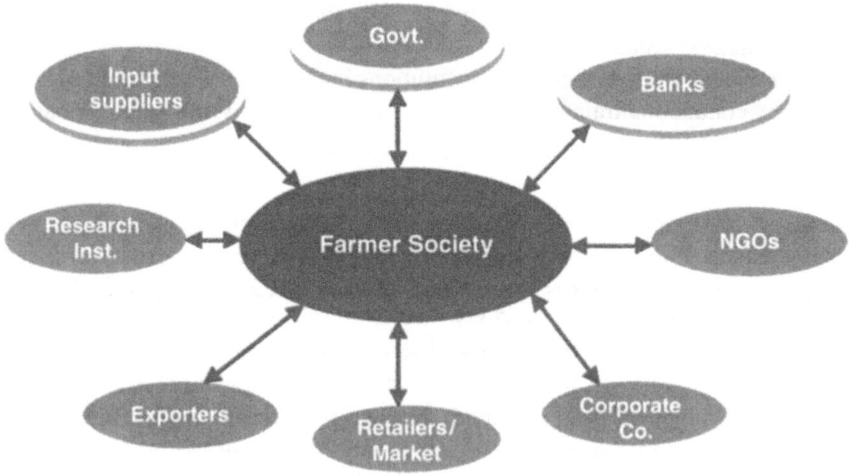

Fig. 3.7 NaCSA model of linking societies to all the key stakeholders

any new disease outbreak, it is easier to coordinate the quick flow of information and samples from field to research institutes, and report back the outcome of the diagnosis and necessary precautionary measures to farmers. Seeing the efficient implementation of the program, corporate companies are showing interest to adopt societies for further infrastructure development. Retailers from developed countries are showing keen interest in buying society produce, which are now becoming increasingly well-known for implementing sustainable farming practices in a socially responsible manner.

NaCSA is working toward developing societies as a potential business model through public-private partnership where all the concerned stakeholders help the societies to sustain for mutual benefit. The success of the approach largely depends on getting market recognition for the traceable, good quality, BMP shrimp produced responsibly and ethically without harming the environment. With the help of all the stakeholders, NaCSA is working towards this goal, thereby making society formation an attractive proposition that can motivate the rest of the farmers to organize themselves into societies. Smallholder farmers will benefit much from increased institutional coordination, capacity, and support in assisting them in managing their societies. They will also benefit from more widespread application of proven technological, social, economic, and governance innovations to reverse losses and improve livelihoods.

3.10 Self Propagating Nature of the Model

Indeed, today we are seeing the emergence of numerous farmer societies throughout India because of the farmers' confidence in participatory group farming and the concept becoming mainstreamed. Cluster organization is progressing very well, which is mainly due to the belief among the farming community that if they have to succeed as shrimp farmers, they have to organize themselves. The reasons for this confidence in group farming according to a farmer are, "*we are strong as a group, we can address issues affecting us, but alone we cannot progress especially in shrimp farming.*" Grassroots action in India demonstrates that group activities of the farmers can improve their well-being in many ways that individual approaches cannot. Farmer organization as groups is generating improvements for the individual producers in the following areas:

- Enhancing their incomes, self-respect, and bargaining power in markets. Clusters offer economies of scale, buying inputs and improve market power when dealing with suppliers and selling product.
- Clusters improve information exchange and sharing of experience among participants.
- Farming skills and technical knowledge.
- Ability to articulate demands and interact with markets and market forces, other political, and social actors.
- Access to financial services and ability to manage funds.

- Knowledge and tools to use information on markets, services, technologies, and rights.
- Self respect, social esteem, and relationships to authorities and other social actors.

With better informed farmers and the spreading awareness building about the society concept, more and more farmers are approaching NaCSA to help themselves organize as societies.

3.11 Cost Benefit Analysis of MPEDA-NACA Project

Tables 3.2 and 3.3 summarize some economic data from 2006, which show that the investment in the project gave a significant economic return. Further economic analysis of the investment would be useful, but the outcomes suggest such investment in small scale shrimp aquaculture is extremely viable from an economic perspective.

Table 3.2 Production summary of BMP and non-BMP farms from 2004 to 2006

	2004		2005		2006	
Parameter	BMP	Non-BMP	BMP	Non-BMP	BMP	Non-BMP
Number of farmers	130		736	425	730	741
Number of ponds	254	187	1187	517	1,370	949
Area (ha)	173	131	663	500	813	605
Production (tons)	40		672	296	870	620
Expenditures/ha (Rs.)	39,141	30,428	173,820	114,899	187,646	133,131
Revenue/ha (Rs.)	48,776	36,651	196,145	119,489	289,229	141,716
Profit/ha (Rs.)	9,636	6,224	22,325	4,590	101,583	19,577

Table 3.3 Project cost, increased profits of BMP farmers and return[a] per unit investment by the program, 2004–2006

Year	Project cost	Increase in profit	Return[a]
2004	1,300,000	590,276	0.45
2005	1,800,000	11,758,305	6.50
2006	3,191,500	66,670,878	20.89
Total	4,991,500	79,019,459	

All values are Indian Rs. (one US$=Rs. 48)

3.11.1 Economic Analysis MPEDA-NACA Project Costs, 2004–2006

1. Cost of MPEDA-NACA project from 2004–2006 is Rs. 4,991,500.
2. Improvement in profit made by BMP farmers compared to non-BMP farmers from 2004–2006 is Rs. 79,019,459.

For each one rupee investment in the program, it generated an average of Rs. 15.8 higher profit for BMP farmers.

3.12 Summary of Positive Impacts

Farmer societies rely more on the fundamental disciplines of sanitation, animal health, nutrition, food safety, and sound management. The BMP implementation through cluster concept has reduced disease risk and made a significant improvement in yields, less impact on the environment, production of wholesome products, and better relations among the players throughout the market chain. In short, it is

Table 3.4 Summary of MPEDA/NACA project impacts

Positive impacts	Remarks
• Reduced disease incidence	• 27% decrease of disease prevalence in BMP ponds compared to non-BMP ponds
• Reduced chemical and antibiotic use and complete traceability of the product	• 10% random BT samples from society ponds tested negative for presence of antibiotics
• Increased opportunity for market access	• Efficiently managed small farmer societies provide similar advantage of integrated larger units • Traceable shrimp from societies. Traceability from brood stock to pond level
• Improved profits • Opportunity for bank credit access	• By reducing the cost of production, profits have been increased. Non-BMP ponds got Rs. 39 for every Rs. 1,000 spent and where as BMP ponds got Rs. 128 for the same amount of investment
• Democratic and transparent societies:	• Democratically organized farmer groups, regular information sharing among farmers
• Sharing of costs	• Cooperation in selecting/testing and buying quality seed and other inputs
• Increased communication	• Farmer field days helping farmers to share their successful experience
• Harmony among farmers	• Each society is having minimum of ten meetings during the crop period
• Lower stocking densities	• The stocking density of shrimp ponds in societies vary from 2 to 6 shrimp per square meter which is far below the level when compared to other countries
• Reduced pollution	• Two societies have adopted organic aquaculture practices
• Increasing awareness on environment	• Abandoned shrimp ponds being revived

helping small farmers to sustain their livelihood through responsible shrimp farming. Specifically, they show evidence of the advantages of small farmers being organized, coming out to mainstream and sharing resources, helping each other and adopting BMPs. The implementation of the BMPs through the cluster concept has provided benefits to the farmers, the environment, and the local community. The summary impact of the project is highlighted in Table 3.4.

3.13 Summary of Lessons Learned

Needless to state in an exercise such as the one presented here that there are many lessons to be learned (Table 3.5). Such lessons are not only useful in improving the processes with time, but could have relevance and application to the development of small scale practices elsewhere, irrespective of the commodities farmed. The lessons learned as work progressed were not only of a technical nature, but also of a significant quantum of it was developing human-relationships, especially between the extension workers, authorities, suppliers, buyers, and the farmers. The latter

Table 3.5 A summary of the lessons learnt over the last 5 years during the success of adoption of BMPs by small scale shrimp farmers and the benefits gained

- Improved farm management practices can reduce environmental impacts, ensure food safety and improve farm profit. The "win-win" situation created by adoption of better management provides a strong incentive for positive change
- Organization of small-scale aquaculture farmers brings about positive social and economic benefits to members. These benefits include
  - Collective planning and shared responsibility helps achieve better management of risks
  - Cluster model of BMP implementation is developing into a self propagating model (farmers believe farmers)
  - Interaction between technical service providers and farmers at the ground level on a regular basis enhances the capacity of both
- Farmer groups can have stronger negotiation power with the input suppliers and traders
- The following points should be considered while organizing farmer groups
  Farmer groups comprise farmers with different needs, interests, skills and financial and technical capacity. A few common interests can hold them together in a group
- Provision of technical services should be independent and without conflict of interest to secure the confidence of farmers
- Investment in institutions (NaCSA) that is focused on small scale farmers can facilitate formation of groups and adoption of BMPs
- It should be recognized that this support takes time, investments in capacity building and institutions are necessary for sustainability
- Revival of the shrimp sector is possible. Shrimp farming can be a source of sustainable livelihood for small scale farmers provided risks are managed through improved management and institution building
- Experiences from India are more widely applicable in other countries across the region
- Links between farmer groups and markets have proved difficult to establish and need to be improved

aspects enable not only easy dissemination of technical information, but also an appreciation of farmer innovations, which are often not taken into consideration. Indeed, in the current day and age, sustainability of small scale farming practices, the backbone of Asian aquaculture, will rest significantly on the successful adoption of farmer innovations.

3.14 Way Forward

For the small scale shrimp farmers to continue to advance, we need a new approach to development. Similarly, for poor and marginalized farmer groups to access benefits of poverty reduction efforts, the position of the farmers in relation to public and private institutions has to change. The farmers must move from being passive recipients of information, services, and regulations to a situation where they take full responsibility for their own development and use public and private institutions as resource providers.

Effectively, engaging with the thousands of aquaculture producers in India and helping them to develop farm level plans for sustainable development will not be a small task, but it is one that can only be achieved with the involvement and contribution of the many players involved in the supply chain, from producers to consumers. The farmers, especially in the current market economies, need the strength that groups can offer for their economic and social advancement. Linking small farmer clusters to sustainability conscious buyers will go a long way in sustaining small farmers' livelihoods.

In this direction, NaCSA is in negotiation with one of the prominent buyers of seafood from USA for purchasing. Society produced shrimp and selling the same with a unique brand name, thereby giving a premium price to the product, which would motivate the farmers to grow the shrimp to the buyer specifications and ensure steady supply of best quality, chemical free, traceable shrimp. This market recognition for the society produce will help us to spread the message of "sustainable aquaculture" far and wide to more areas across India, and will help in sustaining shrimp sector, thereby contributing to a new vision for the aquaculture sector in support of small farmers' livelihoods in India.

Acknowledgment The authors are pleased to acknowledge Mr. G. Mohan Kumar I.A.S. (Chairman, MPEDA) and other staff and officials of MPEDA in Kochi, Andhra Pradesh, Gujarat, Karnataka, Orissa and Tamil Nadu for their kind support and cooperation during the MPEDA-NACA program. The technical services provided by Mr. Mongkhon Primphon, Shrimp farming expert, Thailand, and the cooperation of the Australian Centre for International Agricultural Research (ACIAR) are also gratefully acknowledged. We are also thankful to the All India Shrimp Hatchery Association, Shrimp farmers Association of Andhra Pradesh, Tamil Nadu and Gujarat and Seafood Exporters Association of India for their kind support. The study team is especially thankful to the farmer members of all the Aqua Clubs/societies of Andhra Pradesh, Tamil Nadu, Karnataka, Orissa, and Gujarat for their active participation in the village demonstration program and the MPEDA-NACA and NaCSA team for their dedication toward making the program a success.

References

FAO/NACA/UNEP/WB/WWF. 2006. *International principles for responsible shrimp farming*, 20. Bangkok: Network of Aquaculture Centres in Asia-Pacific (NACA).

Mohan, C.V., M.J. Phillips, B.V. Bhat, N.R. Umesh, and P.A. Padiyar. 2008. Farm level plans/ husbandry measures, Changing trends in managing aquatic animal disease emergencies: Tools for preparedness and response. *OIE Scientific and Technical Review* 27(1).

MPEDA. 2008. News letter. (www.mpeda.com).

MPEDA/NACA. 2003. *Shrimp health management extension manual*. Asia-Pacific: Marine Products Export Development Authority and Network of Aquaculture Centres. http://www.enaca.org/modules/mydownloads/singlefile.php?cid=23&lid=58.

MPEDA-NACA. 2004. *Report on village demonstration programme*. Asia-Pacific: Marine Products Export Development Authority and Network of Aquaculture Centres.

MPEDA-NACA. 2005. *Report on technical assistance on shrimp disease control and coastal management report on village demonstration programme*. Asia-Pacific: Marine Products Export Development Authority and Network of Aquaculture Centres.

Padiyar, P.A. 2005. Contract hatchery systems: A practical approach to procure quality seeds for aquaclubs of small-scale shrimp farmers in India. *Aquaculture Asia* 10(3): 30–31.

Vijayakumaran, M. 1998. Environmental implications of disease treatments in aquaculture. In *Bioethics in India*, eds. Jayapaul Azariah, Hilda Azariah, and D. R.J. Macer, 210–216. Eubios Ethics Institute.

Chapter 4
Backyard Hatcheries and Small Scale Shrimp and Prawn Farming in Thailand

Hassanai Kongkeo and F. Brian Davy

Abstract Thailand has continued to retain the global dominance in shrimp production for over a decade in spite of many adversaries, providing a source of major income, foreign exchange generation, and livelihood opportunities. The Thai shrimp farming sector essentially consists of small scale owner-managed and operated practices, with an average farm size of 1.6 ha. The farming systems have been resilient and adaptive, which has been a key to their sustainability. One of the keys to continued success has been the emergence of backyard hatcheries that provide reliable quality seed stock to the industry. The government support to these hatcheries in the early stages, together with the effective dissemination of culture technologies through the initiatives of the farmers themselves at all stages of the cycle, has facilitated and enabled the farmers to be on a firm footing and encouraged them to embrace changes, and make it sustainable. All these factors together have made the Thai shrimp farming sector a success, while the sector in many of the neighboring countries became almost complete disreputable.

4.1 Historical Development of Backyard Hatcheries

Backyard hatcheries refer to small scale usually family-owned and operated seed production operations that were most often located in the backyard of the owners; hence, the choice of this name (see Plate 4.1).

This close proximity to the family home was critical as it provided family labor, and almost round the clock vigilance, which were the key reasons for success. The development of these hatcheries was closely tied to the overall development of marine shrimp and the giant freshwater prawn, *Macrobrachium*, as well as other cultured marine finfish species, a point that will be dealt with later in this case study.

H. Kongkeo(✉) and F.B. Davy
Netwests of Aquaculture Centres in Asia-Pacific, PO Box 1040, Kasetsart Post Office, Bargkok 10903, Thailand
e-mail: Hassanai.kongkeo@enaca.org

S.S. De Silva and F.B. Davy (eds.), *Success Stories in Asian Aquaculture*,

Plate 4.1 Backyard hatchery

These backyard hatcheries were first developed as part of the major extension thrust in a Thai Department of Fisheries (DOF) – FAO freshwater prawn project that started in 1974. This project was developing technical know-how, but ended up providing free *Macrobrachium rosenbergii* postlarvae (PLs or seed) for stocking for the initial grow-out operations in neighboring farms. Large quantities of PLs were produced at the fisheries station in Bangpakong District, Chacheongsao Province, and distributed by road and rail all over Thailand. However, more interestingly, many of the technical staff of the Bangpakong station also began to produce PLs at their nearby homes, adapting the technology to less conventional smaller holding containers locally used for storing drinking water. Before long, many of the stilted houses of the staff and their friends and neighbors had small production units underneath their living quarters. Even nonscientific staff learned the necessary techniques quickly and then this technology was transferred from DOF technical staff directly to interested farmers. Soon, the first "commercial backyard hatcheries" began to spring up in nearby areas all around Chacheongsao Province. This new approach followed the principle of traditional backyard kitchen garden and fish pond operations, and developed in direct contrast to the massive species-specific concrete fixed structure hatcheries that were set up elsewhere in the 1970s and 1980s. These simple backyard hatcheries were later easily and cheaply modified to produce marine shrimp PLs (*Penaeus monodon, P. vannamei*), grouper (*Epinephalus* spp.) and seabass (*Lates calcarifer*) fingerlings, and other marine finfish species.

This backyard hatchery concept was revolutionary in terms of the low initial investment on land, physical facility construction, and limited need for large scale costly equipment; also the operation costs were relatively low because of the simple techniques and small scale of the operation. These innovations demonstrated the resourcefulness of small farmers and their abilities to take ideas for development and modify them to suit the local conditions. Farmers were eager to experiment and continue learning from their own mistakes. For instance, the use of hypersaline water from coastal salt farms was found to be more cost effective when transported by truck and subsequently diluted to the desired salinity with disinfected freshwater. More importantly, this hypersaline water is pathogen-free (i.e. virus free). The business is organized with these hatcheries purchasing *P. monodon* or *P. vannamei* nauplii from nauplii producers who are located near the open sea areas (better water quality and circulation assist in the maturation process). For *Macrobrachium*, hatchery operators use spawners both from grow-out farms and from the wild. The present success of Thailand in shrimp and prawn industry is testimony to the persistence and ingenuity of Thai farmers in utilizing applied science to its utmost potential.

As mentioned earlier, small hatcheries run by family owner-operators are usually more efficient than larger scale hatcheries, which are operated using hired labor. This family manpower is more flexible with mainly family members, such as the husband, wife or children assisting as needed after school hours. Backyard hatcheries originally started as a secondary occupation for rice farmers or fishers, but soon these activities yielded more income than the primary occupation. The decrease in price of shrimp fry caused by the spread of these backyard hatcheries also helped to stimulate the rapid expansion of grow-out ponds. This family business approach contrasts sharply with the large scale high capital cost hatcheries in which the high fixed costs of wages, power supply, supporting facilities, and other overheads made closure, even for relatively short periods, very difficult.

Backyard hatcheries, in contrast, could discontinue production when disease or other serious problems occurred, even for relatively long periods, without undue hardship when small scale farmers could switch back to their primary occupations of rice farming or fishing. It turned out that periodic discontinuation of the backyard operations was, in fact, useful in checking disease by facilitating the reconditioning, drying, and disinfection of tanks, ponds, aeration, and water systems. These concepts are summarized in the Farm Performance Survey (ADB/NACA 1997), in which Thai small-scale operations yielded more benefits than medium and large scale operations in Indonesia, Taiwan, and Philippines, respectively (Table 4.1).

To complement the backyard hatcheries operations, a variety of ancillary locally developed farm equipment and supply operators developed as well (such as aerators or paddle wheels, water pumps, harvesting sets, commercial larval and adult feeds, probiotic products, and miscellaneous farm supplies). Associated with this was the development of a variety of ancillary businesses such as broodstock suppliers, nauplii producers, PLs distribution, pond preparation, PCR (a disease diagnosis tool) and water quality testing services, harvesting, pond construction, and supply of heavy machines for pond construction. These ancillary businesses developed as different operators identified new entry points in the previously large scale operations.

Table 4.1 Annual financial performance in 1 ha intensive pond (in US$; adopted from Kongkeo, 1997)

Parameter	Indonesia	Philippines	Taiwan	Thailand
Average farm size (ha)	5	9	3	2
Average scale of operation	Medium	Large	Medium	Small
Stocking (PL/m^2)	78	38	73	114
Yield (ton)	6.06	3.05	2.88	10.49
Shrimp sale/kg	6.5	7.1	12.46	6.94
Total shrimp sales	39,390	21,655	35.885	72,801
Labor/kg (% of total)	0.20	0.43	0.20	0.19
	(5.7)	(6.3)	(2.8)	(4.4)
Feed/kg (% of total)	1.41	2.62	1.65	2.01
	(39.9)	(38.4)	(22.6)	(47.1)
Seed/kg (% of total)	0.58	1.27	0.87	0.59
	(16.4)	(18.6)	(11.9)	(13.8)
Power/kg (% of total)	0.36	0.29	0.67	0.33
	(10.3)	(4.2)	(9.1)	(7.8)
Other/kg (% of total)	0.18	0.08	0.58	0.26
	(5.1)	(1.1)	(8.0)	(6.2)
Overhead/kg (% of total)	0.13	0.00	0.68	0.37
	(3.7)	(0.1)	(9.4)	(8.6)
Depreciation/kg (% of total)	0.67	2.14	2.64	0.52
	(18.9)	(31.3)	(36.2)	(12.1)
Production costs/kg (total)	3.53 (21,370)	6.83 (20,820)	7.29 (20,990)	4.27 (44,870)
Net profit margin/kg (total)	2.97 (18,030)	0.27 (880)	5.17 (14,910)	2.67 (27,930)

The risks in these ancillary businesses are also reduced due to shorter periods of operation and the specialized expertise in each business. In addition to the socio-economic benefits to these small-scale operators, local communities were shown to have less social conflicts within their own communities, unlike the experiences of larger scale investments in South America and elsewhere. This is similar to the success of small-scale intensive grow out ponds, which spread all over the country (more than 80% of Thai marine shrimp production came from approximately 12,500 intensive farms in which small farmers typically operate 1–2 ponds with average farm size of 1.6 ha and a total production area of 27,000 ha, Kongkeo 1995, 1997). However, it is recognized that large scale operators are usually important to pioneer development and adaptation of new technologies from government or overseas operations. After being developed in Thailand, the backyard hatchery technology has been transferred through assistance of FAO, Network of Aquaculture Centres in Asia-Pacific (NACA), UNDP, Royal Thai government, and the private sector to Indonesia, Vietnam, India, Bangladesh, and Myanmar. These transfers were then locally adapted, for instance, in some countries, direct seawater was used because they have better seawater supply sources.

4.2 Present Status of Backyard Hatcheries in Thailand

There are more than 2,000 small scale hatcheries in Thailand, mainly located in Chacheongsao, Chonburi, and Phuket provinces, with a total production of more than 80 billion marine shrimp PLs per year (approximately 90% of the total production). Despite their survival over many crises in the past 20 years, these operators recently suffered from disease-related competition with specific pathogen free (SPF) PLs supplied by large scale hatcheries. Such hatcheries can utilize high cost technologies such as SPF and disease resistant strains, bio-secure systems, raceways, etc. taken from overseas. To cover their high investment costs, these large scale hatcheries are under pressure to increase their margin by selling PLs directly to grow out farms, instead of selling nauplii to backyard hatcheries as they had formerly done. After lengthy negotiations, the regulations were then relaxed by releasing SPF nauplii to some backyard hatcheries, which are able to adapt this technology to their present business conditions in 2008. It is expected that the backyard hatchery operators will continue to adapt the SPF concept, but the new world economic crisis caused by US economic slump has led the Thai shrimp industry to reduce its production by 30% in 2008. On the other hand, the Thai shrimp industry is also reducing its dependence on export income through an increase in local consumption, which has reached nearly 35% of the production.

Other new challenges are developing. Traceability of broodstock and certification requirements of developed countries are also difficult problems for these backyard hatcheries because they purchase nauplii from external suppliers. Although nauplii producers can issue PCR negative certificates, it is difficult for them to sort out the source of shrimp (*P. monodon*) by particular backyard hatcheries. Distributors usually mix nauplii from various sources for easy distribution and economic reasons.

The Thai DOF has tried very hard to assist small scale operators. A farm registration system and code of conduct (CoC) and good aquaculture practice (GAP) certification systems have been implemented since 2003. At the moment, 98 hatcheries and 727 backyard hatcheries have been certified with CoC and GAP standards, respectively. Furthermore, the use of the "Movement Document" and traceability system at the grow-out farm level has been recently implemented and is expected to be functioning properly to cover the hatcheries in the next few years.

4.3 History of Freshwater Prawn Farming

As mentioned above, backyard hatcheries grew in relation to freshwater prawn and shrimp farming and it is important to understand their evolution. Besides, both prawn and shrimp farming are dominated by small scale operators and illustrate the related successes of the backyard hatcheries (Kongkeo and New 2008). One of the important milestones in freshwater prawn farming occurred in the late 1970s when the United Nations Development Program decided to fund a 3-year FAO-executed project, "Expansion of Freshwater Prawn Farming," in Thailand (New 2000). As a result of these efforts, farmed freshwater prawn production expanded from less than 5 tons/year before the project began (1976) to an estimated 400 tons by the time it ended in 1981 (Boonyaratpalin and Vorasayan 1983). Soon afterwards (1984), the DOF reported to FAO that Thai production had exceeded 3,000 tons/year (FAO 1989), a very rapid expansion indeed.

This DOF-FAO project not only enabled the establishment of a significant aquaculture sector in Thailand, but also facilitated the development of freshwater prawn farming globally. This facilitation included steps such as the publication of a technical manual on the topic (New and Singholka 1985; New 2002) that was translated into many languages. In addition, the Thai DOF hosted "Giant Prawn 1980," the first international aquaculture conference ever held in Thailand (New 1982). Many Thai experts later advised *Macrobrachium* projects and ventures elsewhere in Asia. By 2005, the aquaculture production of *M. rosenbergii* in Thailand had risen to 30,000 tons/year (valued at US $79 million) and to more than 205,000 tons/year globally (FAO 2007).

Though there has been a great potential for expansion of farming areas and yield, Thailand has maintained production at this level, despite the lack of export markets, unlike for marine shrimp. Similar to mud crab, *Macrobrachium* has a thick shell and a big carapace; when frozen with its shell-on or head-on, it is not well accepted in the international market. Also, it deteriorates after defrosting or refreezing and has a small meat to body weight ratio making it uneconomic for export in the form of frozen peeled prawn. In fact, the head is the most delicious part of prawn for Asian dishes. However, prices in domestic markets are more stable than marine shrimp, which mainly relies on export markets. Similar to the export market of marine shrimp, the domestic market of freshwater prawn was enlarged by reducing the selling price (Fig. 4.1). Farmers also practice partial harvest for the live prawn market to yield better prices. This partial harvest could reduce biomass density, thereby improving the growth of the remaining prawns and reducing the risk of a long culture period. Though there were the serious problems in terms of deteriorated broodstock and Nodavirus outbreak in 2005, farmers could recover in the next year due to their skills in pond management and their efficient scale of operation (Fig. 4.1). In addition, a similar quantity of a related species, *M. nipponense*, was produced in China in 2007. In total, the global farm-gate value of freshwater prawn farming had reached almost US $1.84 billion/year by 2007.

4.4 History of Marine Shrimp Farming

At the beginning, extensive culture of banana shrimp (*P. merguiensis*) and greasy shrimp (*Metapenaeus* spp.) using wild seed stocks were practiced over

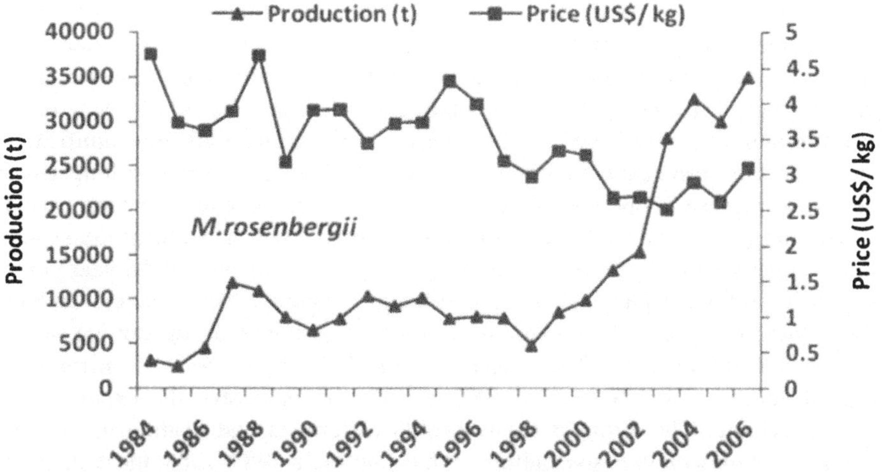

Fig. 4.1 Giant freshwater prawn (*Macrobrachium rosenbergii*) production and average farm gate prices in Thailand (FAO Fishstat Plus)

the past 60 years. Though there was no seawater available, the Bangkok Marine Laboratory successfully cultured to PLs stage *P. merguiensis*, *P. semisulcatus*, *P. latissulcatus*, *M. monoceros* and *M. intermedius* in 1972 (Cook 1973). Seawater had to be brought from offshore by boat. All gravid female shrimp were captured in the Gulf of Thailand. Experiments on pond culture of artificially bred seed were carried out at private shrimp farms in Samutsakorn Province and Bangpoo, Samutprakarn Province, but the results were not satisfactory.

In 1973, the Phuket Coastal Fisheries Research and Development Centre (former Phuket Marine Fisheries Station) successfully bred *P. monodon* by induced spawning from broodstock caught from Andaman Sea. PLs of the early batches were stocked in semi-intensive ponds in Bangkrachai, Chantaburi Province, Klongdaan, Samutprakarn Province, and Klongsahakorn, Samutsakorn Province in 1974. This brought shrimp farming the much needed technique that enabled the farmers to have better control of their crop and sustainable production, instead of reliance only on wild seed for stocking in an extensive culture system. This important research later led to the highest peak of *P. monodon* production of 304,988 m in 2000 (Kongkeo 2006) before substitution by *P. vannamei* and the increased demand for the backyard hatcheries.

Though commercial semi-intensive culture of *P. monodon* using hatchery produced fry commenced in 1974, it expanded rather slowly compared to Taiwan due to the lack of suitable feed, low demand in the internal market (only US $2.50–3.20/kg), and suitable farming practices. Prior to 1985, Taiwan was the only supplier to the Japanese shrimp market due to the cheaper cost of transportation. However, Japan had to stock Taiwanese shrimp in their cold storages for whole year-round consumption. After 1985, the labor and electricity costs went up very significantly due to the Japanese economic boom. Japanese cold storages could no longer bear the heavy increase in operational costs if shrimp had to be stored for long periods (2–10 months in a year). As a temperate country, Taiwan could produce only one crop per year and export to Japan only during a few months of the year before winter. Therefore, Japan had to urge tropical countries, like Thailand and the Philippines, to produce *P. monodon* for a continuous supply throughout the year to save on cold storage costs. Thus, Japan increased the buying price of shrimp from tropical countries to US $8.00–10.00, in order to encourage the expansion of shrimp farming. This brought heavy profits to farmers and leading to the first boom of *P. monodon* intensive culture in the country. In 1987–1988, the collapse of the shrimp industry in Taiwan led to further increases in shrimp production in Thailand.

Similarly in 1993, *P. chinensis* shrimp crops collapsed, probably due to Whitespot viral disease in China; Thailand was able to rapidly increase production to more than 200,000 t to make up for the shortfall in the world supply. The sharp increase in shrimp prices in 1993 (Fig. 4.2) was driven mainly by the high demand in the global market, again spurring shrimp farmers to boost their production.

The outbreak of Yellow-head virus disease in Thailand starting in 1990 did not lower overall production (Fig. 4.2) due to good management practices and efficient small scale operations. Whitespot disease outbreak slightly affected the production during 1994–1997, before launching again in 1998, because the technologies on intensive closed system and on-screening of broodstock and PLs by PCR test were well adapted by farmers. These technologies were simplified and locally adapted to suit local conditions and farmer capabilities.

Similarly in 1993, *P. chinensis* shrimp crops collapsed, probably due to whitespot viral infection. In 1999, the Thai Government issued a decree to ban *P. monodon* farming in inland areas to avoid the negative environmental impact from saline water, which was not fully enforced until 2001 (Fig. 4.2).

Consequently, 30% of the total shrimp farming area that was located in freshwater areas of the country was affected by this decree. Farmers could not revert back to *Macrobrachium* culture, which yielded a very low price and had a limited market. Therefore, SPF *P. vannamei* from Hawaii was first introduced by the private sector in late 1999 for trials to replace *P. monodon* in freshwater areas. In fact, research has been conducted successfully in China to culture *P. vannamei* in freshwater areas without any introduction of saline water. After successful trials in Nakorn Pathom farms, it spread throughout all inland shrimp farming areas within 2 years. However, culture in pure freshwater yielded only small shrimp (15–20 g) and mortality occurred after 3 months. Because of the small size, farm gate price was low at the beginning of this development (Fig. 4.3), but this was compensated

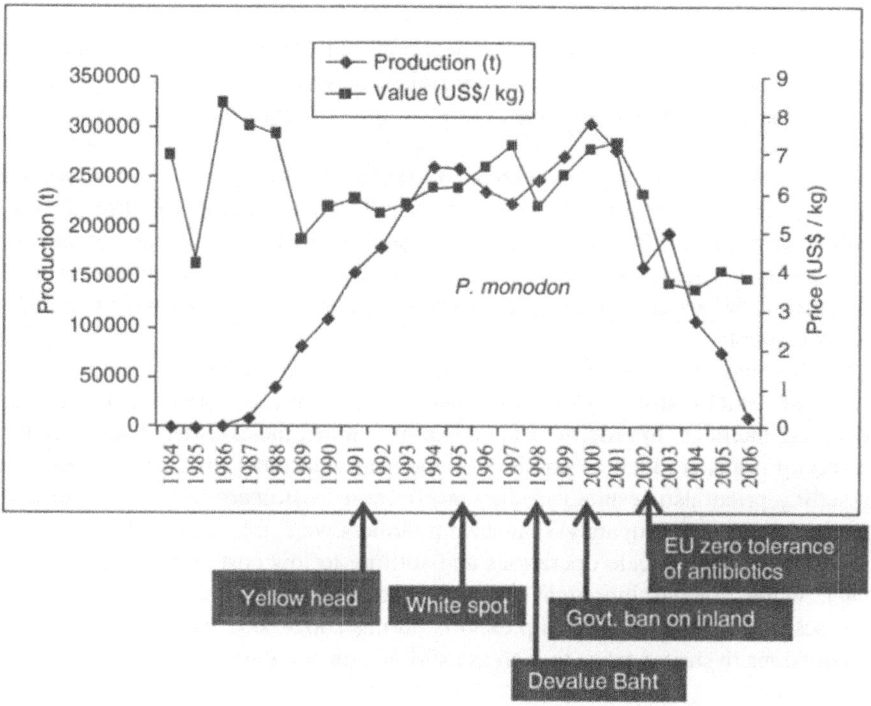

Fig. 4.2 Giant tiger prawn (*P. monodon*) production and farm gate price in Thailand (FAO Fishstat Plus and CP Shrimp Culture Newsletter)

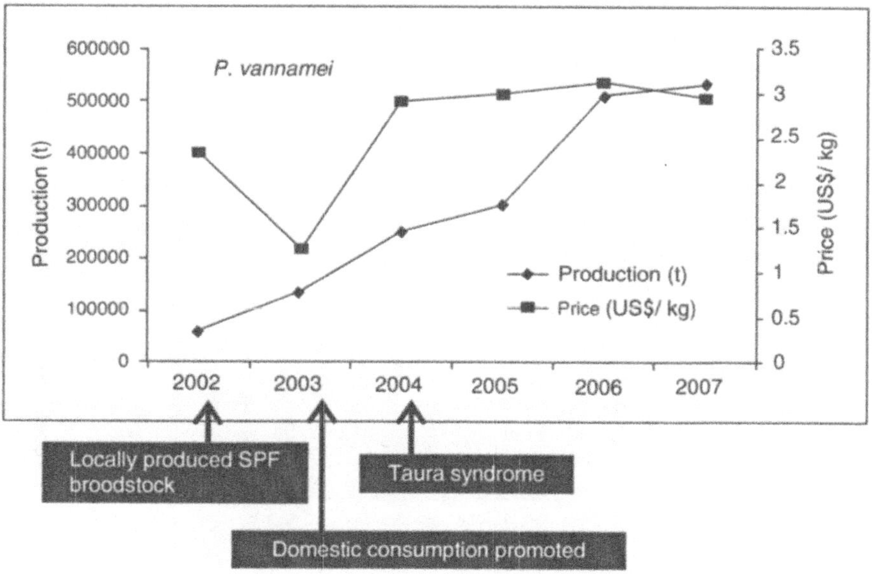

Fig. 4.3 Pacific white shrimp (*P. vannamei*) production and average farm gate price in Thailand (FAO Fishstat Plus and CP Shrimp Culture Newsletter)

by increasing the number of crops, for example, to 3/year. When *P. vannamei* culture expanded into coastal areas beginning in 2003, production increased sharply. Prawn size and farm gate price also increased up to 20–30 g and US $3.0–4.0/kg, respectively, because saline water prolongs the culture period to 3.5–4.5 months.

Similar to *P. monodon*, an outbreak of Taura syndrome virus in *P. vannamei* began in 2004, but it did not lower the overall annual production (Fig. 4.3) as locally produced SPF PLs were commonly used for stocking. Government and the Thai Shrimp Farmers Association also promoted domestic consumption (as now more than 30% of total production) to avoid export price drops, which resulted from oversupply.

Thailand has been able to keep its position as a world leader in shrimp exports since 1988 (FAO Fishstat) due to its constant supply of raw material and efficient processing facilities. In order to expand export volume, the exporters had to reduce the shrimp price as much as possible to compete with other producers. Lowering the selling price also resulted in drawing in more customers in both export and domestic markets. Fortunately Thai shrimp farmers were able to reduce production costs by using small-scale operations and shifting to low cost species like *P. vannamei,* while maintaining high standards and quality of their products. Sharp increases in the volume of shrimp exports during 2002–2007 were mainly caused by reductions in shrimp price to increase shrimp consumption (Fig. 4.4). This trend

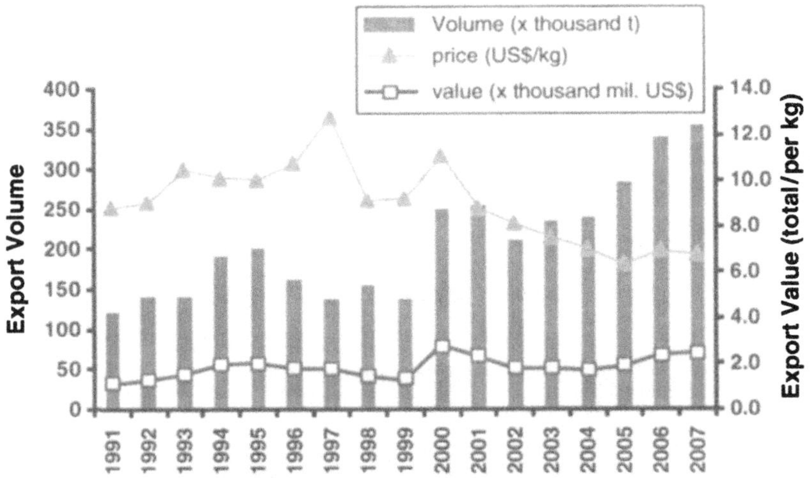

Fig. 4.4 Thai shrimp export volume, value and FOB price (Royal Thai Department of Customs)

mirror imaged the trends of the past with commodities such as salmon and chicken.

4.5 Key Factors for the Success of Small Scale Operators

In fact, the success of backyard hatcheries and small scale shrimp/prawn farmers more generally must be seen as part of a dynamic process involving continual adaptation to solve new problems. To be the leader in global shrimp production is not easy, but to keep this position for more than 20 years is even more difficult. These small scale farmers played a very significant role in maintaining this coveted status. They were able to survive during crises through which large scale operators could not. The key factors for their success include the following.

4.5.1 Key People and Organizations

Farmers always have new ideas for development or modification, and are eager to experiment on their own. This is similar to the success of fruit, rice, and orchid farming in Thailand. Due to the long experience in aquaculture, particularly freshwater species and extensive marine shrimp farming, Thai farmers are able to cope very well with advanced technologies. Shrimp culture technology was mainly transferred among farmers from lead farmers who were the more innovative initial risk takers. It has been proven that shrimp culture technologies were rapidly transferred by observation and enthusiastic learning mostly from more experienced farmers, than through the government extension services. For instance, DOF budget for information and training is only 7.2% of total DOF budget (Table 4.2) or 0.38% of total shrimp ex-farm value, or 0.30% of total export value.

Nevertheless, government has an important role in areas, such as in research and development, regulating the industry, improving infrastructure (e.g. water sources), as well as providing many free services (disease diagnosis, diatom stocks, water and

Table 4.2 Budgetary allocations of the Thai Department of Fisheries (DOF) 2008

Item	US $	%
Research and development	15,332,024	18
Information and training	6,155,827	7.2
Rural aquaculture	4,935,960	5.8
Quality control	26,926,481	31.7
Fisheries control	8,721,464	10.3
Stock enhancement and resource improvement	22,602,196	26.6
Other	305,411	0.4
Total	84,979,363	100

soil analyses, shrimp consultancy, and market advice). DOF Quality Control Section has a larger budgetary allocation (32%) and plays a very significant role in improving the seafood quality (international food safety standards and requirements from individual importing countries). Not only the DOF, but also the Ministry of Commerce, assists the shrimp industry in the promotion of shrimp products in domestic and international markets, improving the image of shrimp culture in international forums and negotiating trade barrier issues with importing countries. Universities and vocational colleges also assist in capacity building of resource persons for the shrimp farming industry. Many skilled farmers, who became leaders of farmer groups, were later trained and received further education in aquaculture. Many universities also conduct high quality research on *P. vannamei* and *Macrobrachium* farming in Kasetsart University, domestication of *P. monodon* and PCR in Mahidol University, shrimp genetic and recycling pond in Burapa University, and treatment of farm effluent and sludge in Prince of Songkhla University.

Feed factories also play important roles in shrimp extension programs through their sales services, including provision of regular training programs at their training centers through the support of overseas training for selected farmers.

4.6 Government Policy and Support

As Thailand is an agricultural country, government has had a major policy to promote crop production as "the world kitchen" and fully aims for development to assist the poor small scale operator. These policies directly benefit over 30,000 families of small scale farms, over 2,000 backyard hatchery families, and also many thousands of laborers in 184 shrimp processing plants, 49 canneries, and over 10 shrimp feed mills (DOF 2007). Good infrastructure facilities, such as roads, electricity and water, were critically important initial facilities provided to these areas. Free technical assistance through DOF is the main policy that assists small scale farmers. DOF policies for disease prevention also permitted and strictly regulated only the SPF *P. vannamei* from Hawaii to enter the country. This partial relaxation on the import of certified, SPF PLs from Hawaii prevented the private sector and other interested parties from importing PLs from elsewhere (which could have been viral contaminated, thereby leading to serious impacts on the industry in Thailand in a major way). As mentioned earlier, the Ministry of Commerce, in cooperation with DOF, heavily promoted sales in both domestic and international markets. Price stabilization policies also helped small scale farmers to sell their products with guaranteed prices offered through the Agricultural Bank. The Agricultural Bank usually provides loans with minimal interest rate only to small scale farmers. Government also provided income tax exemption schemes to small scale farmers and fishermen if their net profits do not reach the ceiling.

4.6.1 Networking and Information Exchange

Networking and Information exchange have been very important mechanisms for farmers to avoid/reduce their losses from disease outbreak, drops in market price, increase in feed and seed prices, poor weather condition, etc. Information on new initiatives or shrimp farming products, alternate species, and advanced technology also help farmers to improve their profits. Such information will support them to make judgments regarding shifting cultured species, farming practices, or temporary discontinuation of production.

At the village level, farmers generally exchange their ideas on new initiatives, market information, disease outbreaks and aquaculture news with neighboring farms through visits and/or via mobile phone. Farms are always open for visitors who need information even without any appointment. In some villages, they also have established shrimp farmer clubs as a forum for regular meetings to exchange ideas and information, as well as to increase the bargaining power with shrimp buyers/collectors. These farmers' clubs were also established at district and provincial levels for the same purposes. At national level, shrimp farming associations and shrimp industry association (farms, feed plants, processing plants, and exporters) were established mainly for policy making and planning for shrimp culture development, information exchange, coordinating with government or outsiders, harmonizing trade conflicts, and bargaining with importers. In addition, information is exchanged at the regional or international level through regional organizations, such as the NACA, South East Asian Fisheries Development Centre (SEAFDEC), INFOFISH, shrimp website, international/regional magazines, and conferences.

Farmers also receive information on shrimp markets and new technologies from newsletters/publications from major shrimp feed producers, the national shrimp farming association, local shrimp farmer club, DOF magazine, over ten local aquaculture magazines, as well as the internet. Such information is also exchanged during meetings or conferences regularly organized by national associations, local farmer clubs, universities, DOF, feed producers, processing plants, and the direct communication with shrimp buyers/collectors and processing plants.

4.6.2 Crisis as One Key Driver of Success

As described above, there have been a number of crises (mainly diseases and market changes), which have acted as barriers to the success, but more importantly served as key drivers for innovations by small scale operators. These crises included:

- Yellow-head disease outbreak in *P. monodon* beginning in 1990
- Whitespot disease outbreak in *P. monodon* starting in 1994

- Devaluation of the Thai currency (Baht) in 1997, which caused significantly price increases in imported artemia, artificial plankton, chemicals and drugs, oil, etc.
- Government ban on *P. monodon* farming in inland areas in 1990, which made farmers switch to *P. vannamei* culture
- Strict application of EU regulations on zero tolerance for antibiotic residues in 2001
- Taura syndrome outbreak in *P. vannamei* beginning in 2004
- Tsunami disaster which damaged many hatcheries and broodstock centers in the south in late 2004
- Limited distribution of SPF nauplii from major commercial SPF broodstock centers to backyard hatcheries in 2007
- Traceability and certification requirements by importing countries in 2007
- Global economic downturn caused by US economy and increase in oil price in 2008

4.7 Switching of Crustacean Species and Sustainable Farming

P. vannamei is not native to Asia (Fig. 4.5) and was first introduced to Taiwan in early 1990s to replace the problematic species, *P. monodon*. It was later transferred to China, Thailand, and Indonesia in early 2000s. Asia accounts for 85% of *P. vannamei* global production in comparison to its native countries in Eastern Pacific (15%). Though it has been introduced for culture on the Atlantic side of Central and South America, Tahiti, Hawaii, and Taiwan nearly 20 years back, it is not significantly observed in the wild. This may be due to its poor mating habit, which is the major problem in hatchery production.

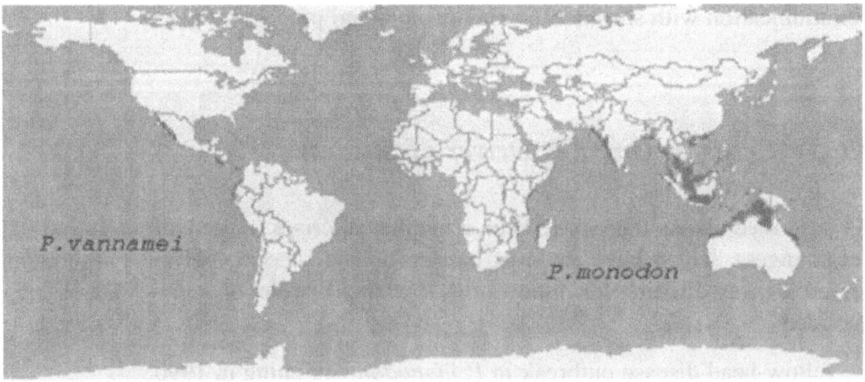

Fig. 4.5 Geographic distribution of native shrimp species, *P. monodon* and *P. vannamei* (Holthuis 1980)

In principle, for culture of any crop, farmers should have alternate species for their sustainable production, livelihoods, and profits. They should have a wider range of species for selection when the existing species encounters problems such as market, diseases, environment, etc.

Increases in *P. vannamei* production in Asia are mainly due to:

- Shifting from problematic species, *P. monodon* in Southeast Asia and Taiwan as well as *P. chinensis* in China
- Its ability to be domesticated and use as SPF broodstock compared to *P. monodon*;
- Less virulent diseases
- Tolerance to wider ranges of salinity and temperature
- Better survival in poor pond bottom condition according to its schooling habit
- Simple hatchery and grow-out technologies
- Lower production costs, particularly seed, feed, water exchange, and aeration

In fact, reduction in the use of wild *P. monodon* broodstock may improve its health and enable a rebuilding of the wild population for future aquaculture use.

During its development in Asia, the industry had to solve the problems of price drops in international markets mainly due to oversupply. Thus, both Chinese and Thai industries launched heavy promotion for domestic consumption and intra-regional trade. As a result, domestic consumption in China and Thailand reached 90% and 30%, respectively, in 2005. They also reduced the cost of production by using lower protein diet, higher stocking density, less water exchange, etc.

New diseases such as Taura syndrome, Infectious Hypodermal and Hematopoietic Necrosis (IHHN) were also introduced to the region. Thai farmers learned from the more virulent white spot and yellow head diseases of *P. monodon* how to solve the problems from these new diseases. For instance, SPF PLs from locally adapted broodstock are commonly used by farmers. Closed system small ponds (<1.0 ha) were another simple innovation to reduce stress by starting with an initial low salinity environment.

There are still many constraints that need to be solved by shrimp industry for their sustainability in the future such as:

- Increasing costs of aeration, water pumping, and transportation
- Declining demand: less consumption of seafood (as a luxury commodity) following the world economic crisis
- Climate change is bringing a variety of impacts (warmer surface water, increased frequency and intensity of storms, flood and cold bottom currents affecting wild broodstock)
- Drops in supply of high protein fishmeal
- New disease outbreaks leading to long-term sustainability of farms;
- More export competition if the rapid growth of Chinese economy slows leading to oversupply to the domestic market
- Under threat in key export markets due to adverse publicity concerning environmental impact such as mangrove, saline intrusion, etc.
- Increasingly stringent international rules and agreements, such as
  - Food safety and trans-boundary movement

Table 4.3 Shrimp production from CP Royal Shrimp-cum-rice Project, Prachinburi

Parameter	*P. monodon*	*P. vannamei*
Salinity (ppt)	2–10	2–10
Density (PL/m^2)	27	60
Culture period (days)	135	120
Survival rate (%)	72	83
Harvest size (g)	26.3	16.6
Production (kg/ha)	5,117	8,281

- Eco-labeling and traceability
- Biodiversity concerns regarding the introduction of exotic species
- Trade barriers and specific rules (e.g. GSP in EU, antidumping and C-bond in USA, GMOs in soybean, child labor, human trade, animal welfare, other social parameters, etc.)

In general, the international environment is continuing to present new challenges with the increasing international rules for the Asian shrimp industry. Small farmer operations face a variety of challenges here, many of which may only be solved by stronger networking and social organization. In early 2008, rice prices increased dramatically due to the global shortage of food and conversion of food crops to energy crops. Therefore, many *Macrobrachium* and *P. vannamei* farmers in freshwater areas in Thailand converted to the more profitable rice farming in their ponds. Organic load left from previous shrimp crops provided fertilizer for rice cultivation saving inorganic fertilizer costs. Similarly, many marine shrimp farms in brackish water areas also converted their ponds to rice production in the rainy season and switched back to shrimp production in the dry season. Before rice planting, farmers wash the surface of the pond bottom with freshwater to remove precipitated salt. Periodical discontinuation of shrimp production and replacement by rice cultivation will also break shrimp disease cycles and improve the quality of the deteriorated pond bottom (see the average results of 15 shrimp crops in Table 4.3).

References

ADB/NACA. 1997. *Shrimp and carp aquaculture sustainability*. Proceedings of Regional Study and Workshop on Aquaculture Sustainability and the Environment, Beijing, China, October 1995. Bangkok, Thailand: NACA.

Boonyaratpalin, M., and P. Vorasayan. 1983. Brief note on the state of the art of Macrobrachium culture in Thailand. NACA Working Paper WP/83/7. Bangkok, Thailand: NACA.

Cook, H.L. 1973. FAO Report to the Government of Thailand on Shrimp Farm Development. FAO Report No TA 314. Rome: FAO.

DOF. 2007. Fisheries Statistics of Thailand, Department of Fisheries Circular No. 6/2007, Bangkok.

FAO. 1989. Aquaculture Production 1984–1986. FAO Fisheries Circular, 815. Rome: FAO.

Holthuis, L.B. 1980. Shrimps and Prawns of the World. An Annotated Catalogue of Species of Interest to Fisheries. FAO Fisheries Synopsis No. 125, FAO Catalogue vol. 1.

Kongkeo, H. 1995. How Thailand made it to the top. *INFOFISH International* 1/95: 25–31.

Kongkeo, H. 1997. Comparison of intensive shrimp farming systems. *Aquaculture Research* 28: 789–796.

Kongkeo, H. 2006. *Responsible shrimp farming: A critical overview.* Presentation in WAS'06 Conference, Florence, Italy.

Kongkeo, H. and M.B. New. 2008. The successful development of backyard hatcheries for crustaceans in Thailand. *Aquaculture Asia* 13(1): 8–11.

New, M.B. ed. 1982. *Giant Prawn Farming. Developments in Aquaculture and Fisheries Science,* 10. Amsterdam, The Netherlands: Elsevier.

New, M.B. 2000. History and global status of freshwater prawn farming. In *Freshwater prawn culture*, ed. M.B. New and W.C. Valenti, 1–11. Oxford: Blackwell Science.

New, M.B. 2002. Farming freshwater prawns: A manual for the culture of the giant river prawn (*Macrobrachium rosenbergii*). *FAO Fisheries Technical Paper*, 428 [Also published in Mandarin, with Arabic, French, Malayalam and Spanish versions in preparation].

New, M.B., and S. Singholka. 1985. Freshwater prawn farming: A manual for the culture of *Macrobrachium rosenbergii. FAO Fisheries Technical Paper*, 225, Rev. 1 [Also published in Farsi, French, Hindi, Spanish & Vietnamese].

Chapter 5
Cage Fish Culture: An Alternative Livelihood Option for Communities Displaced by Reservoir Impoundment in Kulekhani, Nepal

Tek B. Gurung, Raja M. Mulmi, K.C. Kalyan, G. Wagle, Gagan B. Pradhan, K. Upadhayaya, and Ash K. Rai

Abstract In Nepal, fisheries and aquaculture were hardly envisaged during the planning of the hydropower project inception phase of the impoundment of Kulekhani Reservoir. However, one of the joint projects between the International Development Research Centre (IDRC), Canada and the Government of Nepal demonstrated that cage fish cultivation in the reservoir is a promising alternative livelihood option for displaced communities. Impoundment of these riverine waters left most aboriginal communities who had been living in the valleys subjected to impoundment, depriving them of livelihood assets they have enjoyed for generations. However, adoption of plankton-based cage fish culture has demonstrated that such displaced communities can be successful in fish farming as an alternative livelihood opportunity. Such "no feed" farming systems seem to fit more appropriately into the needs of the farmers living around Kulekhani, and offer an increased potential for long-term sustainability. Among 500 families displaced in 1982 due to impoundment, nearly 81% adopted cage fish farming in the reservoir. A total of 231 families are now organized into 11 farmer groups as part of an amalgamated fish farmers' association. Most recent data show that the total fish production from the reservoir approximates 165 tons (2005/2006 data), of which 130 tons were from cage culture. Production has been ongoing for a period of 21 years, and all these communities have been able to generate a higher income, which consequently have resulted in significant improvements to their livelihoods leading to associated benefits such as sustaining the schooling of their children from primary to university levels. Data indicate that livelihoods could be further improved if a sustainable fingerlings supply mechanism could be developed in Kulekhani. It is also important to note that over the years cage fish farming in the reservoir also stimulated the development of a capture fishery, based on escapees and naturally recruited species, all of which have significantly contributed to increasing a fresh affordable animal protein source to the nearby communities.

T. B. Gurung(✉)
Fisheries Research Division, Godawari, PO Box 13342, Kathmandu, Nepal
e-mail: frdgdr@wlink.com.np

S.S. De Silva and F.B. Davy (eds.), *Success Stories in Asian Aquaculture*,

5.1 Introduction

Inland fishery is a time old tradition of most early civilizations. However, in recent times, inland waters have been increasingly impounded for hydropower generation, irrigation, and other purposes (McCully 1995; De Silva 1998), with devastating impacts on nearby agrarian and other communities (IEA 2000; IRN 2003). Impounding submerges the plains suitable for human settlement, agriculture, and several other uses, affecting the traditional livelihoods of these communities dependent on those lands. River impoundment is known to bring about severe and acute social problems in hilly areas where flat land is scarce.

The impoundment might include involuntary resettlement that often may not be managed fairly, leading to human tragedies such as uprooting and dispersal of communities (IEA 2000). There are several examples around the world where displaced communities have lost all their livelihood resources, even after receiving heavy subsidies and compensation provided for resettlement (Cernea 1997; Ghimire 2004). This might be associated with the sudden turnover, as often people fail to accept sharp changes to their ongoing livelihood strategies. The process of resettlement is a daunting task. Nevertheless, many governments are convinced that such projects with good resettlement programs can generate collective benefits, in terms of generating electrical power or other uses, which largely exceed the adverse impacts (Pradhan 1987; IEA 2000).

Elsewhere in the region it has been shown that reservoir-based fisheries and aquaculture have successfully generated food, income, and alternative livelihood opportunities (Costa-Pierce and Hadikusumah 1990; Sugunan 1995; Costa-Pierce 1998; Gurung et al. 2008). However, there are limited studies explicitly explaining the role of fisheries and aquaculture on the issue of resettlement of communities displaced from impoundment.

Here, we present a case study where ethnic aboriginal communities displaced by impoundment of Kulekhani Reservoir for hydropower generation adopted cage fish culture as a successful strategy for an alternative livelihood option. Thus, the objective of this presentation is to examine the impact of cage fish culture on resettlement of the displaced during the construction of hydropower dam, to evaluate the present status of cage fish culture and its level of production, and to discuss further measures for scaling up fish production in Kulekhani Reservoir.

5.2 Cage Fish Culture

Cage fish culture is considered to be an old tradition that has developed into a major sector in aquaculture only in the recent past (De Silva and Phillips 2007; Tacon and Halwart 2007). In Nepal, fish culture in floating net cages commenced only around 1971 (Sharma 1979).

The characteristic feature of the Nepalese cage fish culture is its sustainable production through the use of plankton-feeding fish, such as bighead carp, *Aristichthys nobilis,* and silver carp, *Hypophthalmichthys molitrix* (Swar and Gurung 1988; Rai and Mulmi 1992, Gurung et al. 2005). In this system, stocking is managed at a level of sustainability at which fish production remains low per unit area compared to intensive cage fish farming as the fish rely only on natural food without external feed input (Rai and Mulmi 1992). Approximately 3–5 g body weight fingerlings of silver carp and bighead carp are stocked in nursery net cages of 5×5×2 m of mesh size of 5–10 mm, grown for about 8–12 months, and transferred to production cages of 25–50 mm mesh cages of 5×5×2 m, hung on bamboo frames that also function as a float. The stocking density ranges from 10 to 15 fish m^{-3} depending on the size of the fish. This method is extensive, and therefore, an environmental friendly farming system as it does not introduce nutrients into the water body.

5.2.1 Origins of the Idea of Cage Fish Culture and its Expansion in the Reservoir

The plankton-based cage fish culture in Nepal was initiated in lakes of the Pokhara Valley for the benefit of the landless and deprived *Jalari* community for livelihood enhancement (Gurung and Bista 2003). As this cage fish farming practice was successful (Swar and Gurung 1988; Gurung et al. 2005), it was thought that cage fish farming could also provide an alternative livelihood for displaced people in Kulekhani Reservoir. A proposal from the Government of Nepal was accepted by the International Development Research Centre (IDRC), Canada, to research on limnology and assess the feasibility of developing cage fish farming in Kulekhani Reservoir (Rai and Mulmi 1992; De Silva 1998). Based on the findings that the technology was feasible, appropriate cage fish culture was subsequently extended successfully into Kulekhani reservoir (Rai and Wagle 2007).

5.3 Kulekhani Reservoir

Kulekhani Reservoir (27°23′25″–27°41′31″N; 85°2′46″–85°16′16″E) that was located approximately 42 km from the capital Kathmandu in the northeastern part of Makawanpur District with a catchment of 126 km^2 was impounded in 1982. Kulekhani Reservoir is a small 220 ha water body situated in the mid hills of Central Nepal at 1,430 m above sea level. The reservoir was designed with an anticipated lifespan of more than 50 years. The reservoir draws down upto 54 m during the peak season of power generation, reducing the water surface area down to about 65 ha. Earlier, it was assumed that most of the time the water level in the reservoir would remain stable because the main aim of the reservoir was to supply hydropower generated electricity only in case of emergency. However, this was not to be so. At present, there has been a huge power shortage facing several hours of load shedding in Kathmandu. As a result, the reservoir has been used intensively to fulfill the power requirements, causing occasional severe reduction in reservoir water level.

Fisheries and aquaculture were not considered at the inception of hydropower projects in Nepal, such as the impoundment of Kulekhani Reservoir. However, reservoir fisheries in many countries have contributed considerably to national fish production and livelihood generation (De Silva and Phillips 2007; Tacon and Halwart 2007). Thus, initiation of reservoir fisheries could offer a significant contribution to fish production and socioeconomic development of local communities. In view of this, IDRC (Canada) and the Government of Nepal jointly operated an inland fisheries project in the reservoir through which cage fish culture was initiated. Cage fish culture activities have subsequently been expanded, providing an alternative livelihood option for the communities displaced by the impoundment.

Information on cage fish farming was collected in the present study, primarily by a Rapid Assessment Survey in July 2008. The past Annual Reports published by the Fisheries Development Office were analyzed to obtain basic water quality parameters, cage fish production, fingerling supply mechanisms, marketing, new trends, and other socio-economic factors. During the study, 53 farmers, members representing 11 farmer groups including the President of Kulekhani Fish Farmers' Association, were interviewed.

5.4 Displaced Ethnic Communities and Their Settlement

The total population of the Kulekhani watershed was 36,187 in 1981 and comprised *Brahmin, Chhetri, Magar, Newa, Tamang,* and other ethnic communities (Bjønness 1983). Among them, 500 families or approximately 2,100 people were affected during the construction of the hydro-dam in Kulekhani. According to the Planning Commission, 60% of the population in the area was classified as poor, out of which 36% were living below the poverty line (Bjønness 1983).

5.4.1 Cage Fish Farming

Cage fish culture in Kulekhani Reservoir can be divided into government and private sectors. In the government sector, only a limited number of cages, maximum 90 in 2001/2002 with a total volume of 4,584 m^3 is used (Table 5.1).

The main aim of the government component was to monitor the productivity of the reservoir in terms of fish production and support fingerling distribution services to farmers. The total caged fish production ranged from 1.5 to 5.7 tons in different

Table 5.1 Cage volume, fish production, and productivity estimates of government sector in Kulekhani Reservoir

Year	Nursery cages		Production cages		Total		Production (kg)
	No.	Vol. (m³)	No.	Vol. (m³)	No.	Vol. (m³)	
97/98	11	503	39	1925	50	2428	1501
98/99	11	503	52	2462	63	2965	3089
99/00	13	672	61	2874	74	3546	4058
00/01	13	672	63	3180	76	3850	5717
01/02	16	812	74	3772	90	4584	4530
04/05	12	507	48	2024	60	3045	4330
05/06	12	507	48	2024	60	3045	4700
06/07	12	507	48	2024	60	3045	4060

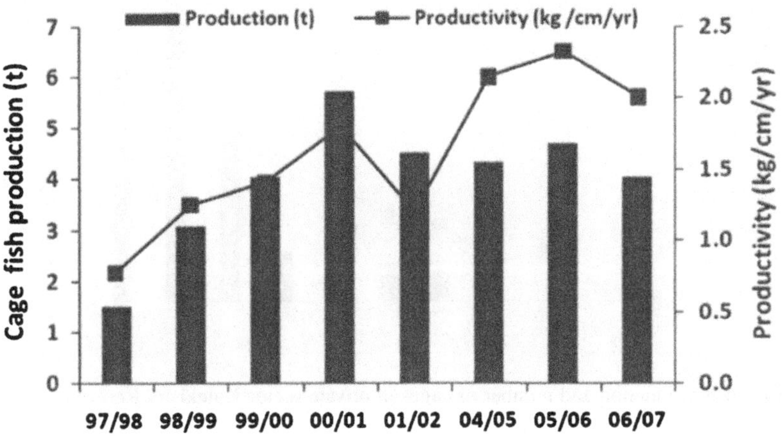

Fig. 5.1 Fish production and productivity estimates from government sector production cages

years, with a production ranging from 0.78 to 2.32 kg m^{-3} during 1997/1998–2006/2007 (Fig. 5.1).

5.4.2 Private Sector Cage and Capture Fisheries

Private sector cages are owned and managed for fish production by individual farmers themselves, with support from the government on provision of fingerlings (1–3 g) of bighead carp and silver carp as major species. Common carp and grass carp were also distributed occasionally.

Over the years, the number of cages in the reservoir and the cage fish production increased gradually and significantly (Fig. 5.2). With the increase in cage numbers, a capture fishery began to be developed based on escapees augmented by naturally recruited species. Consequently, the total fish production from the reservoir approximated 165 tons in 2005/2006, while in 1997/1998, it was only 5.5 tons, primarily from government and private cages and recapture of fish stocked in the reservoir (Fig. 5.3). This represents a substantial increase of 30 times in only 8 years. Of the total production in 2005/2006, cage fish culture contributed 79% with the remainder from open water recapture fisheries (Fig. 5.3).

About 60 fishers and farmers are engaged in capturing indigenous and exotic fish from open water in the reservoir and its inlet streams, using gill nets, cast net, and other means. Most fishers also owned cages in the reservoir. In the surrounding

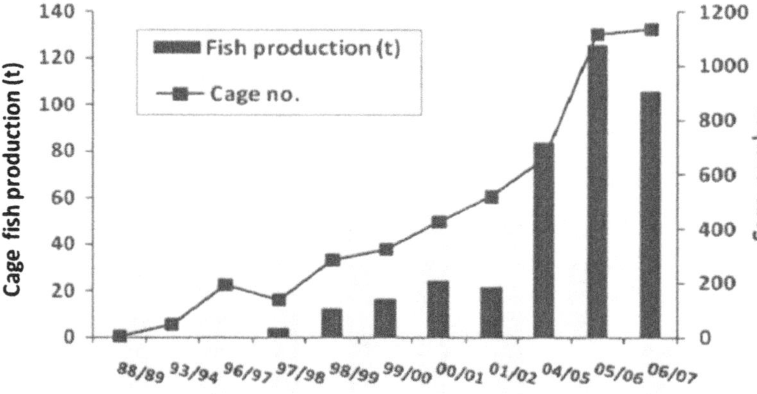

Fig. 5.2 Fish production and number of cages in private sector Kulekhani Reservoir

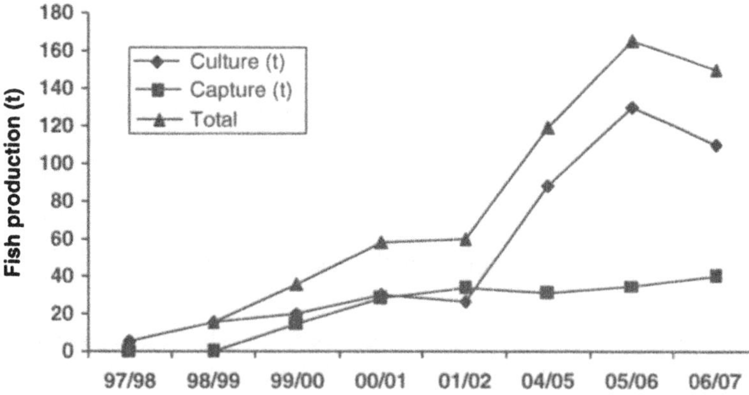

Fig. 5.3 Trends in total, cage and capture fishery production in Kulekhani Reservoir

small hill streams and rivulets, Asla (*Schizothorax* spp.) and Katle (*Neolissocheilus hexagonolepis*) are caught as the major native species, while exotic carps, such as bighead, silver, and grass carp, were also caught in the reservoir. To increase the fish yield from the open water capture fisheries, a restocking program was initiated with fingerlings from both private and government sectors. The maximum production from the open water recaptured fisheries in the reservoir and surrounding incoming streams reached 40 tons, approximately 26% of total fish production from the reservoir in 2006/2007 (Fig. 5.3).

Fingerling supply: Cage fish culture in the reservoir was dependent from the start on a fingerling supply from outside of Kulekhani as a fish hatchery has not been developed for fingerling production for the main fish species, bighead, and silver carp. Fingerlings were supplied from different government run hatcheries in the country, though the supply of fingerlings was reported to be often inadequate. Over the years, the number of fingerlings supplied for cage culture has increased steadily, reaching 774,000 in 2005/2006 from less than 200,000 in 2000/2001. To fulfill the increasing fingerling demand, farmers themselves purchased and transported fingerlings from different locations within the country, as well as from India, mainly Kolkata. However, farmers lack special fingerling transporting vehicles and had to rely on public transportation, such as buses, trucks, and minibuses.

Fish production: In the cages, mainly bighead and silver carp were raised in polyculture with their stocked ratio depending on the availability of fingerlings. Common carp, grass carp, and other species were also captured from open water. The performance of rainbow trout (*Oncorhynchus mykiss*) maintained on a pelleted feed has recently been tested in cages (Bista et al. 2004).

Marketing: A local market close to the eastern side of the reservoir has developed to market fish produced from the reservoir. The market is supplied with caged and captured fish from the reservoir for sale to local and people traveling from Hetauda-Kathmandu and vice versa. There are about five fish retailers in the area who are directly in contact with groups of farmers to purchase fish with cash directly. Retailers, however, complained of insufficient availability of captured and cultured fish, but cage owners do not have problems of marketing either captured or cultured fish. All fish from the reservoir were mostly sold fresh. The marketing of cage and net material was an important issue as farmers wished to buy durable and high quality net materials that would last for a longer time span of 15–20 years.

Socio-economics of cage fish culture: Fifty three cage fish farmers were questioned on their origin, profession, and problems encountered with their livelihoods. Of these, 81% represented family members of communities displaced during the impoundment of the reservoir. According to Bjønness (1983), nearly 500 households were displaced and lost their house, agricultural land, and watermill during the impoundment of the reservoir. The proportion of ethnic groups displaced were *Newar* (41%), *Tamang* (33.3%), *Magar* (10. 3%), *Chettri* (7.7%), *Brahmin* (5.1%), and *Kami* (2.6%). Communities belonging to *Tamang*, *Brahmin* and *Chhetri*, *Newar*, *Magar* and *Kami*, *Damai*, and *Poude* ethnic groups owned cages and were engaged in cage fish culture since its inception. Cage ownership according to ethnic groups were *Tamang*, *Magar*, *Newar*, *Brahmin* plus *Chhetri*, and *Kami* were 31.8, 29.2, 20, 17.9, and 0.9%, respectively.

To promote cage fish culture, the local community Fisheries Development Center, Kulekhani, Watershed Conservation Department and Local Development Office contributed substantially to different social activities. In addition, the local Fisheries Development Center used short-term project support provided under Hill Agriculture Research Council. Plan International Makwanpur also assisted the financially deprived groups of the communities. Other supporters included the Institute of Agriculture and Animal Sciences (IAAS), financed by the National Agriculture Research and Development Fund, Fisheries Research Station, Pokhara, Integrated Bagmati Watershed Project, Agriculture Development Bank and Local Micro Finance Companies. These agencies organized training, interaction meetings, visits, and farmer to farmer programs with the aim of capacity enhancement of farmers engaged in cage fish farming.

Farmer Associations: At present, there are 11 cage fish farming groups in Kulekhani and they have formed the "Indrasarobar Fish Growers Association," Markhu (Table 5.2). Under this association are involved 231 families, among which 111 farmers are women and the remaining 120 are male members. The farmers groups were established informally around 1993, and formally registered in 1998 in the District Administrative Office with the objective of a collective approach to promote cage fish farming and initiation of a robust marketing channel of fish. The main reasons for the formation of the groups and Association were social security, control over poaching and conflict resolution, and to provide a "social shield" to cage fish culture activities. In this connection, the Association has contributed towards managing watchmen's houses in the reservoir, where representatives of the Association keep vigilance of poachers. An agreement was reached that only 1,000 m^3 of cage could be owned by an individual farmer, though it was not implemented for many years as the Association was inactive. All members in the group need to pay a membership fee.

Table 5.2 Details of fish farmers groups in Kulekahani Reservoir

Group name	Village	Members
Dhaneshwor Matsya Palan Krishak Samooh	Markhu	25
Lail Gurans Matsya Palan Samooh	Markhu	25
Kali Devi Matsya Palan Krishak Samooh	Tuamath	14
Sungabha Matasya Palan Krishak Samooh	Kulakhani	13
Annapurna Matsya Palan Samooh	Markhu	13
Sateswar Matsya Palan Samooh	Markhu	20
Sri Ganesh Matsya Palan Krishak Samooh	Markhu	22
Saya Patri Matsya Palan Krishak Samooh	Kulekahni	11
Nayagaun Matsya Matsya Palan Sammooh	Kulekahni	13
Namuna Matsya Palan Krishak Samooh	Kulekahni	10
Milijuli Mahila Samooh	Markhu	8

There are several influential farmers operating cage fish culture activities successfully in the Association. There are also some who have not been able to repay the bank loan. Among the most influential farmers, Mr. Ram Prasad Nagarkoti is well known (Box 5.1).

5.5 Cage Net Materials and Accessories

Since ready-made net cages or net materials are not manufactured in Nepal, farmers depend on suppliers/traders for obtaining cage material and accessories. In Pokhara where cage fish culture was initiated first, nylon and polyethylene cages were imported from Japan, Norway, and Britain. Some of these ready-made cages were brought to Kulekhani later, but when the demand for cage materials increased, local dealers in Kathmandu imported the net material from China. Such cage material and accessories may cost up to NRs 8,000–10,000 (US $ 126–158 @ 63 Rs. to 1 US $) for a single cage of approximately $5 \times 5 \times 2$ m cage with top and bottom covers. These materials are often rated low from life expectancy perspectives and quality of the twine used.

5.5.1 Gill Nets and Cast Nets

Net materials for gill nets were generally brought from India or fishers might buy the net materials from local traders. The price of single gill nets depends on the length of the net and ranges from NRs 1,000 to 2,000 (US $ 15–30). Similarly,

Box 5.1 Mr. Ram Prasad Nagarkoti (46): The Case of a Successful Cage Fish Farmer in Kulekhani Reservoir

Mr. Nagarkoti, 46 year old, is one of the successful cage fish farmers in Kulekhani Reservoir. He initiated cage fish culture with few cages in 1993/1994 with the support of Fisheries Development Center, Kulekhani. He himself worked on the cages; cleaning, guarding, and protecting from possible threats and natural disasters. As a result, he achieved better production. Later years with income from the fish sales, he added more cages, and at present, has 53 cages. His annual income has now reached approximately about 0.2 million NRs. With the support of the income generated, Mr. Nagarkoti has built a house in Kathmandu, which is considered an asset in Nepal. He runs a hotel in Kulekhani with the income from the fish sales.

wooden plank boat used for navigation and transportation in everyday work could be bought in Pokhara earlier; however, local carpenters can now also build boats. Such a boat of medium size might cost up to NRs 40,000.

5.5.2 Economics of Cage Aquaculture

Cage fish farming has improved the socioeconomic conditions of the families involved. The cage fish culture technology has expanded enormously in the last 21 years after its initiation in the reservoir (Box 5.2). At present, the total number of cages in the reservoir is upto 1,163 and 72 in private and government sectors, respectively. The total cage volume is 55,981 m^3 (private-52,936 m^3 and government owned 3,045 m^3).

5.6 New Trends

An initial trial to grow rainbow trout (*Onchorynchus mykiss*) in cages in the reservoir has been started recently. During winter, trout can be cultivated in cages; however, the supply of trout fingerlings of desirable size could be a

Box 5.2 A Chronological Order of Important Events in Respect of Adoption and the Success of Cage Fish Farming from the Time of Impoundment of the Kulekhani Reservoir

1982–1984:	Land requisition, dam construction started, (Bjønness 1983)
1983:	Nearly 500 people displaced (Bjønness 1983)
1984:	IDRC and HMG started the work (Bajaj 1998)
1988:	Cage fish culture started by 5 farmers with 125 m^3 in private sector
1992:	Project hand over to Government of Nepal
1993:	Cage No 47, Cage vol. 1974 m^3
1996:	March-April. During intake repair sudden lowering in water level caused severe mortality of fish. NEA gave a compensation of 0.9 million NRs to 36 farmers (Siwakoti 2002)
1999–2000:	Total 92 families, 323 cages, (65 women 27 male) in 11742 m^3
2001:	Plan International Makwanpur supported the supply of cages to 25 females
2002–2004:	Commencement of Hill Agricultural Research Project
2007–2008:	11 farmers' group 223 families (81% displaced communities).

At preset cage, fish culture sustains 41 (45%) primary school kids, 35 (38%) Secondary level, 15 (16%) college

problem at present and needs to be resolved and managed in the future (Bista et al. 2004).

Cage owners have reported occasional mortality in carp. It is speculated that this is due to glass fish (*Chanda* spp.), accidentally introduced from the southern Terai with the transport of fingerlings. This fish is known to remove the scales of silver and bighead carp in the cage (Gurung et al. 2008). Later the open wounds are infected by fungi, causing mortality. To prevent the entry of glass fish into the cages, farmers have started to use smaller mesh size net cages. However, use of smaller mesh cage could reduce the exchange of water containing plankton; this could result in slower growth of caged fish and unwanted deposition of feces and uneaten plankton debris in the bottom of the cage. In general, major fish disease problems have not been encountered so far, except for a few accidental incidents of reservoir drying and harm caused by natural calamities.

5.7 Discussion

The construction of the dam, reservoir, and road affected 321 houses, nearly 2,100 people belonging to 500 families (Bjønness 1983). Among the displaced communities in 1983 (Bjønness 1983), about 31.8, 29.2, 20, 17.9, and 0.9% were from *Tamang, Magar, Brahmin, Chhetri, Newar, Kami, Damai* and *Pode*, respectively, who were involved in cage fish culture activities. The present ethnic makeup of people involved in cage fish culture suggests that indeed a significant proportion of the people who were displaced during the impoundment adopted cage fish culture as a livelihood. Some displaced people migrated to various places and most of them were not able to utilize the compensation for establishing an alternative livelihood, probably due to inability in managing the money as most were illiterate with poor knowledge, skills, and social relations. After the displacement, several families could not recover their old status and prosperity (Bjønness 1983).

It is found that nearly 81% of people displaced during the impoundment have adopted cage fish farming as their alternative livelihood option, which very positively augmented their source of income and means. This implies that cage fish farming provided a very positive alternative approach to their livelihoods. It demonstrates that cage fish culture had been a suitable and manageable technology which can easily be adopted by rural people and fishers. For example, in Pokhara Valley lakes, a landless ethnic traditional fisher community comprising approximately 300 families has successfully adopted the technology of cage fish farming as a livelihood means (Swar and Pradhan 1992; Gurung and Bista 2003; Gurung et al. 2005).

The fisheries including that of cage fish culture have never been part of the resettlement strategies of displaced people among planners and policy makers in

Nepal in the past. However, recently the Nepal Electricity Authority has realized the importance of fisheries promotion. For instance, a fish hatchery has been envisaged to employ the displaced ethnic local fishers to breed and release the fingerlings of native fishes into the upstream and downstream reaches of Kali Gandaki River, where Nepal's largest hydropower dam has been constructed recently. Such attempts are appreciated where traditional fishers can use their knowledge and skills of river fish handling not only for the benefit of themselves, but also for local and global communities, as a mitigation strategy for the maintenance of biodiversity.

At present, there was no system developed to record the production of fish from private cages; however, based on our estimates of productivity from government sector cages in 2005/2006, the annual production of fish from cages was likely to be around 130 tons, sustaining 231 families living on the catchments of the reservoir. Such success in the Kulekhani in the promotion of livelihoods of displaced communities compares well with that in Indonesia, where several thousand displaced people were engaged in fish production using the cage fish culture technology (Costa-Pierce and Hadikusumah 1990; De Silva and Phillips 2007; Tacon and Halwart 2007).

In plankton-based cage fish culture activities, the private sector involvement occurred around 1992 (Rai and Mulmi 1992; De Silva 1998; Rai and Wagle 2007). Since then the involvement of displaced communities in cage fish culture increased substantially in the last 21 years. Most of the members suspected that the low fish production in 2006/2007 was due to overcrowding of the cages in the reservoir. In general, the productivity estimate in the year 2006/2007 was also low compared to previous years. The slow growth of fish into the cage might be due to reduced plankton production in the reservoir, and merits further study.

The trend of fish seed supply to farmers confirms a rapid increase in fish seed demand for cage fish culture in the private sector. This sharp inclination in fingerlings distribution from the government sector confirms the increasing popularity of cage fish culture given increasing number of cages in the reservoir every year. Based on estimates of the optimum number and volume of cages in the reservoir, fish seed needs are suggested to be approximately 600,000 for cage culture only (at the stocking rate of 10 fish m^{-3}).

However, the current fingerling distributed to the private sector is only about 700,000 for all activities including reservoir stocking. Also given the need to transport fingerlings from far distant places, it is easy to speculate that there might be undesirable mortality of fingerlings after stocking. This probably is a major reason behind the farmer complaints of inadequate fingerlings supply for cage fish culture.

Prior to impoundment, the small rivulets and tributaries were well renowned for Asla or snow trout (*Schizothorax* spp.) abundance. In the past, once a week, Asla were sent from here to the Royal Palace in Kathmandu (Bjønness 1983). Recent studies of the Kulekhani area indicated that after the impoundment, the fish species combination in recapture fisheries has changed due to alteration in physical, chemical, and biological changes in water quality (Swar 1992).

Cage operations in the reservoir are managed by the farmer association under a jointly developed set of rules and regulations. The farmer associations were established in 1993, initially involving only a few farmers. At present, there are 11 different associations. These have been unified into a single umbrella organization of cage fish farmer associations comprising about 231 households with 120 men and 111 women members (data up to 2007). The reason for the increasing number of groups and associations seems to be the popularity of cage culture as an assured source of economic returns, and the understanding that the cage farmers need to act collectively than individually.

In 1993, the reservoir was damaged by a devastating flood (Winrock International 2004) when the cages were badly damaged/smashed, but fortunately restored later as part of the reservoir repair. Siwakoti (2002) described how the fish farmers were affected and the ensuing conflict between the fish farmers and Nepal Electricity Authority was finally resolved through a negotiated dialog process. This incident also suggests that cage fish culture might be achieving its rightful recognition as compared to the state of thinking at the time of inception planning for the reservoir. In the future, it seems that the association will provide an important negotiating and pressure group, if undesirable future events occur in Kulekhani.

The nutrient loading in the reservoir from the population in the catchment area is suggested to remain in the reservoir for a long time (Ghimire 2004). Generally, carp in cage fish culture grow slower than in the other parts of the tropics, but given this nutrient loading and temperature, the growth rate may be considered to be moderate. The present trends of total fish production in Kulekhani Reservoir were compared with previously reported studies (Bhukaswan 1980). Indeed the fish productivity in the cage fish cultivation ranged from 0.77 to 2.32 kg m^{-3} $year^{-1}$ in the reservoir. The lowest productivity was in the beginning (1997/1998), but increased in later years with some fluctuation and peaked in 2005. Comparatively, the low productivity might be associated with several factors. The highest water temperature in the reservoir was 25.1°C in July. The water temperature for fish growth remains favorable from March to October; in rest of the year, water temperature is below 18°C (November to February) with the lowest of 12.4°C. The lowest transparency (in August) was due to silt and mud being carried by rainwater due to heavy monsoon rains. However, highest transparency occurs in winter and is due to "clear water phase" and mixing in the water column. The highest dissolved oxygen 7.7 mg L^{-1} occurred in October in surface water, while the lowest 3.3 mg L^{-1} was in June.

5.8 Contribution of Fisheries to Social Development

Well-managed small scale inland fisheries could generate wealth and sustainable economic growth in reservoirs. Such reservoirs have been referred to as "bank in the water" (Béné 2006; Dugan et al. 2007). Dugan et al. (2007) mentioned that fish are nonconsumptive water users and their cultivation could be highly productive; however, planners often give little recognition to freshwater fish production. Some have suggested that inadequate and exaggerated fish production statistics might be some of the reasons for the low priority in policies, leading to the allocation of national development resources away from fisheries (Dugan et al. 2007). The present findings of successful adoption of cage fish culture in support of the livelihoods of displaced communities in Kulekhani Reservoir suggest that the capacity of fisheries was indeed poorly recognized by decision makers as fisheries and aquaculture were not included in the operation plan since the beginning.

Recently, however, the planners have given emphasis on establishment of fish hatchery in hydropower projects. The present experience is suggestive that in future, fisheries research and development planning should be part of hydropower projects for additional social development, hopefully leading to more mainstreamed thinking in this regard. The summary details of this chronicle of cage fish culture development in Kulekhani are given in Box 5.2. Further economic analysis is needed to evaluate, more deeply, how far cage fish culture contributes to the livelihood of displaced communities and the improved understanding of the importance of fisheries in the reservoir.

In Nepal, nearly half of the economically active population (47.8%) were illiterate and about only 25% were just literate (Adhikary 1995). The successful adoption of cage culture farming by displaced communities has also been responsible for major social developments with regard to education of the children in the families (Fig. 5.4). This has been brought about as a result of financial security and family stability, and the community endeavoring to better itself. Almost all the children of the farming community receive some form of education, unlike in many rural communities in Nepal.

Also the proportion receiving secondary and tertiary education is comparable to that observed in urban communities in Nepal.

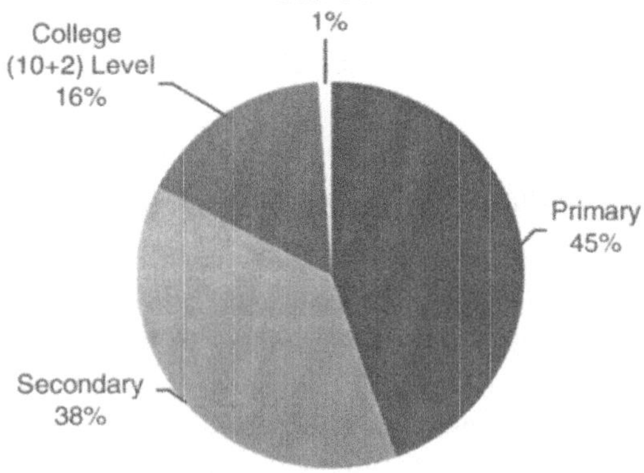

Fig. 5.4 The proportion of farmer community children in the different levels of education

5.9 Contribution to Gender Empowerment

Cage fish culture adopted in Kulekhani has contributed to strengthen women's empowerment. One aspect is the strengthened role of women in the decision making process. Among the farmers involved in cage fish farming, 111 families are primarily represented by women. These women actively take part in all activities concerning fisheries in Kulekhani, like attending meetings, workshops, dealing with fingerling transportation, fish stocking, boating, harvesting, and marketing. One study reported that fisherwomen usually do not consume the captured or cultured fish at home, as fish could be sold at a relatively higher price. Instead they mostly buy water buffalo meat to supplement their household demand for animal protein supply (Gurung 2005). This might be a strategy to save money they earn from fish production. In Nepal, men also help their wives; the increased male activity might lead to an increase in the morale of the women, which is also considered an important aspect of success (Bhujel et al. 2008).

In Nepalese society, generally all household chores are the responsibility of women; however, this work was not considered as "work" (Gurung 2006). Cage fish culture activities in Kulekhani have been successful in allowing women to move away from other household chores and move to more day-by-day decision demanding activities thus contributing to their capacity enhancement in their day-by-day decision making activities in the society.

5.10 Scaling up Strategies

Regardless of the many problems described (IEA 2000; IRN 2003), cage fish culture in Kulekhani Reservoir presents one of the most promising impacts with substantial increases in the number of cages and farmer participation over the last 21 years in Nepal. This cage fish culture technology has set a good example of success to provide livelihood to displaced communities during the impoundment of the reservoir. In 1988, there were only few families involved in cage fish culture. At present, there are nearly 231 families dependent on cage fish culture. It is anticipated that among the various issues suggested (IEA 2000; IRN 2003), if fish farming could be included into the plan of action as one of the mitigatory approaches of damming from the inception phase, it might help to reduce the social, environmental, resettlement, and economic conflict to substantial extent.

Pradhan (1987) has emphasized that for future damming, the present experience of social, economical, and environmental benefits should be scaled up to provide for better welfare and opportunities for utilization of resources created by impoundments. At present, Nepal is facing a critical energy shortage. Neighboring India is facing a similar crisis. These indicate that in the near future, there will be pressure on the use of river water for hydropower dam construction in Nepal. In the near future, at least two reservoir-based hydropower dams are coming up. One is planned in far-western development region in Western Seti (ADB 2007; Gurung et al. 2008) and another in the western development region in Seti River. Both these reservoirs would be situated in areas feasible for carp culture, and in all probability, cage culture could be the best option.

5.11 Conclusions

Cage fish culture in Kulekani has become a promising intervention to improve the livelihood of displaced communities as it offered the best economic options after inundation of their cultivated land. In addition, streams and rivulets around Kulekhani were well renowned fishing grounds before inundation, and as a result, many families were familiar with fish handling, and this might have led to more rapid acceptance of a fisheries-related occupation. It can also be concluded that probably the scarcity of flat land for agriculture around the reservoir and nearby places might have also led to the adoption of the cage fish farming strategies relatively readily.

The success of cage fish culture in a hydropower reservoir, which has not been envisaged at the beginning, suggests that the capacity of fisheries has been poorly understood by decision makers. The present experience also suggests that with the inclusion of fisheries into the integrated action planning additional environmental, economical and social advantages from the hydropower projects can be obtained.

Acknowledgement We are thankful to IDRC project on inland fisheries in Kulekhani, which enlarged to such an extent that it provides livelihood for several hundred people who never had the experience of fish farming before. Special thanks to Prof. Sena De Silva, Director General, Network of Aquaculture Centers in Asia-Pacific (NACA), whose efforts on cage fish culture could contribute as an alternative job opportunity to displaced communities in Kulekhani Reservoir. Our sincere gratitude to all who contributed directly or indirectly to make cage fish culture project a success in Kulekhani. The contribution of Ministry of Agriculture and Cooperative, Department of Agriculture and Nepal Agricultural Research Council is acknowledged. We also thank Nepal Electricity Authority for providing necessary support to cage fish culture program. Our sincere thanks are also extended to all farmers group for their cooperation. Special thanks to Mr. Sahdev Kumar Rimal, the Secretary, Indrasarobar Fish Growers Association, Markhu; Mr. Jay Dev Bista and Mr. Suresh Kumar Wagle Senior Scientist, Fisheries Research Station, Pokhara; and Dr. Madhav Kumar Shrestha, Institute of Agriculture and Animal Sciences, Rampur Campus, Tribhuvan University for their support. Fund for this study was provided by ICEAID.

References

ADB. 2007. Nepal: West Seti hydropower project, Environment assessment report. Prepared by West Seti Hydro Project for the Asian Development Bank, Project No. 40919, 68.

Adhikary, N. 1995. Impact of inland fisheries project on farmers living around indrasarobar reservoir, Kulekhani, Makawanpur District, Nepal. Submitted to International Development Research Centre (IDRC), Canada, 121.

Bajaj, M. 1998. Survey and assessment of IDRC's completed projects: Social policy, public goods and quality of life issues, three case studies, Resource costs of under nutrition and morbidity, India (93–8300), Informal sector Street Food, Pune, India (87-0053), Inland Fisheries (82-01910). ACHIV 613.2 (540:541.35), 36 pp.

Béné, C. 2006. *Small-scale fisheries: Assessing their contribution to rural livelihoods in developing countries*. FAO fisheries circular no. 1008. Rome: Food and Agriculture Organization.

Bhujel, R.C., M. K. Shrestha, J. Pant, and S. Buranrom. 2008. Ethnic women in aquaculture in Nepal. *Development*, 51: 259–264.doi:10.1057/dev. 2008, 11.

Bhukaswan, T. 1980. Management of Asian reservoir fisheries. *FAO Fisheries Technical Paper*, 207, 69 pp.

Bista, J.D., T.B. Gurung, S.K. Wagle, G.B. Thapa, and R.P. Dhakal. 2004. *Cage culture of rainbow trout (Onchorynchus mykiss) in stagnant water in lake and reservoir*. A paper presented in 7th Asian Fisheries Forum held in 30th November 2004 in Hotel Equatorial, Penang, Malaysia, 12.

Bjønness, I-M. 1983. Socio-economic analysis of the effects from the Kulekhani Hydro-electric project Nepal, Research Working Paper, Nepal National Commission for UNESCO, National Committee for the man and Biosphere, Ram Shah Path, GPO Box 1071, Kathmandu, Nepal. University of Oslo, Department of Geography P.O Box 1042, Bilidern, Oslo 3, Norway, 94 pp.

Cernea, M.M. 1997. *Hydropower dams and social impacts: A sociological perspective*, 1–37. Environmentally Sustainable Development (ESD). The World Bank.

Costa-Pierce, B.A. 1998. Constraints to the sustainability of cage aquaculture for resettlement from hydropower dams in Asia: An Indonesian case study. *Journal of Environment and Development* 7: 333–368.

Costa-Pierce, B.A., and H.Y. Hadikusumah. 1990. Research on reservoir-based cage aquaculture systems in Saguling Reservoir, West Java, Indonesia. In *Reservoir Fisheries and Aquaculture Development for Resettlement in Indonesia*, eds. B.A. Costa-Pierce, and O. Soemarwoto, ICLARM, Technical Report 23. Manila: International Centre for Living Aquatic Resources Management (ICLARM).

De Silva, S.S., ed. 1998. *Reservoir fishery management and development in Asia*. Proceedings of a workshop held in Kathmandu Nepal, 23–28 November, 1987. Ottawa: IDRC, xii+246.

De Silva, S.S., and M.J. Phillips. 2007. A review of cage aquaculture: Asia (excluding China). In *Cage aquaculture – Regional reviews and Global Overview*, eds. M. Halwart, D. Soto, and J.R. Arthur, 1–16. *FAO Fisheries Technical Paper*, 498. Rome: FAO.

Dugan, P., V.V. Sugunan, R.L. Welcomme, C. Béné, R.E. Boummett, M.C.M. Beveridge, K. Abban, and U. Amarasinghe, et al. 2007. Inland fisheries and aquaculture. In *Comprehensive assessment of water management in agriculture. Water for food, water for life: A comprehensive assessment of water management in agriculture*, ed. D. Molden, 459–483. London: Earthscan/Colombo: International Water Management Institute, 664. Download: www.fao.org/ag/agl/aglw/aquastat/water_use/index6.stm.

Ghimire, B.N. 2004. *Economic losses to Kulekhani hydropwer project due to siltation in the reservoir. Final report*, Winrock International: Bhubaneshwor, 36 pp.

Gurung, A. 2005. *Dietary pattern, reproductive health and nutritional status of fisherwomen in Pokhara Valley, Nepal*. Master thesis in home science, Padma Kanya Campus, Kathmandu, Nepal.

Gurung, M. 2006. *Women in fisheries: A study on women's coping strategies in fishing communities in Lake Begnas*, Nepal. A thesis submitted in partial fulfilment of the requirement for the Degree of Masters of Science in International Fisheries Managemnet Norwegian college of Fishery Science University of Troms, Norway, May 2006, 47pp. (Source: http://www.nfh.wit.no/dok/manju_gurungo.pdf

Gurung, T.B., and J. D. Bista. 2003. Livelihood improvements through fisheries in the *Pode* community in Pokhara, Nepal. *STREAM Journal Volume* 2 Number, 1–3.

Gurung, T.B., S.K. Wagle, J.D. Bista, R.P. Dhakal, P.L. Joshi, R. Batajoo, P. Adhikari, and A.K. Rai. 2005. Participatory fisheries management for livelihood improvement of fishers in Phewa Lake, Pokhara, Nepal. *Himalayan Journal of Sciences* 3: 47–52.

Gurung, T. B., S.K. Wagle, J.D. Bista, and R.M., Mulmi. 2008. *Reservoir and lake fisheries management in Nepal*, 18. A paper presented in reservoir and Lake Fisheries management project planning meeting. Thailand: NACA-ICEIDA. 13–16 January 2008.

IEA. 2000. *Hydropower and environment: Present context and guidelines for future actions*, International Energy Agency, subtask 5, Volume 2, Main Report, 182 pp.
IRN. 2003. *Beyond hydropower energy options for the 21st century*, 1–2. International Rivers Network, 1847, Berkeley Way, Berkeley, CA 94703, 94703, 510-848-1155. www.IRN.org,1–2.
McCully, P. 1995. *Silenced rivers. The ecology and politics of large dams*, 350. London: Zed Books.
Pradhan, B.R. 1987. *Potential of reservoir fishery development in Nepal*, 29. Proceedings of workshop on reservoir fishery management and development in Asia. Kathmandu: IDRC, 264.
Rai, A.K. and R.M. Mulmi. 1992. Observation of planktivorius fish species in floating cages in Indrasarobar Reservoir. *Journal of Natural History Museum* 13(1–4): 3–9.
Rai, A.K. and S.K. Wagle. 2007. Cage fish culture – successful income generation in manmade reservoir Kulekhani, Markhu, Nepal. *Aquaculture Asia Magazine* 12(2): 1–50.
Sharma, R.N. 1979. Cage fish culture in Nepal. In *International cage and pen fish farming*. 12–21 February Tigbauan, Phillipines: Aquaculture Department SEAFDEC, 93–97.
Siwakoti, G "Chintan," 2002. *Nepal: A case study of Kulekhani Hydropower project and conflict management through dialogue*. Diologue on water, food and environment, A paper presented in International Water Conference, Hanoi, Vietnam, October 14–16, 2002, 3.
Sugunan, V.V. 1995. Reservoir fisheries of India. *FAO Fisheries Technical Paper*, 345, 423 pp.
Swar, D.B. 1992. Effect of impoundment on indigenous fish population in Indrasarobar Reservoir, Nepal. In *Reservoir fisheries of Asia, Proceedings of the second workshop on reservoir fishery management in Asia*, ed. S.S. De Silva, 111–118, Ottawa: IDRC.
Swar, D.B. and T.B. Gurung. 1988. Introduction and cage culture of exotic carps and their impact on fish harvested in Lake Begnas, Nepal. *Hydrobiologia* 166: 277–283.
Swar, D.B. and B.R. Pradhan. 1992. Cage fish culture in lakes of Pokhara Valley, Nepal, and its impact on local fishers. *Asian Fisheries Science* 5: 1–13.
Tacon, A.G.J., and M. Halwart. 2007. Cage aquaculture: A global overview. In *Cage aquaculture – Regional reviews and global overview*, eds. M. Halwart, D. Soto, and J.R. Arthur, 1–16. FAO Fisheries Technical Paper, 498, 241.
T-HPC. 2000. *The Theun-Hinboun Power Company's (T-HPC) Mitigation and Compensation Program*, 50. Theun-Hinboun Power Company Limited, Vientiane, Lao PDR.
Winrock International. 2004. Financial Incentives to Communities for Stewardship of Environmental Resources, *Feasibility Study*, 60. Winrock International 1621, North Kent Street, Suite 1200, Arlington, Virginia, USA.

Chapter 6
Enhancing Rural Farmer Income through Fish Production: Secondary Use of Water Resources in Sri Lanka and Elsewhere

Upali S. Amarasinghe and Thuy T.T. Nguyen

Abstract The inland fishery of Sri Lanka has been essentially a capture fishery from major and medium scale irrigation reservoirs. However, small-sized (<100 ha) minor irrigation reservoirs are also frequent in the country. Naturally, these water bodies are incapable of supporting self-recruiting fisheries, but can be utilized to enhance the fish production significantly through development of culture-based fisheries (CBF), without causing impediment on their primary use.

CBF in village reservoirs of Sri Lanka is a communal activity, and the water bodies used for this purpose are rural, thus benefiting rural communities by augmenting their traditional means of incomes and also increased food-fish availability. CBF is a non-water consumptive secondary activity that brings into play communities that were not engaged in fishery-related activities previously. It is also environmentally friendly, as the only external input is the seed stock. CBF is a present day paradigm of ecosystem-based aquaculture.

CBF has developed as a result of coordinated efforts of a multitude of stakeholders, working in unison, resulting in developing and improving the knowledge base, facilitating required legislative change, such as in the case of amendment of the Agrarian Development Act, and issue of a governmental decree permitting and encouraging CBF in Sri Lanka and community organizations. CBF is a sustainable activity that also impinges on bringing better harmony amongst rural communities. Most importantly, CBF brought about socio-economic benefits to the rural communities.

In a similar vein, CBF activities are being adopted in the mountainous region of Northern Vietnam, and in reservoir coves and flood plain depressions in Lao PDR, with results complimentary to those from Sri Lanka. In all the instances, CBF activities have been sustainable with a strong community-based management strategy driving it. In all the countries, rural farming communities are adopting CBF in suitable water bodies with resulting improved income generation and food-fish availability to the rural communities.

U.S. Amarasinghe (✉)
Department of Zoology, University of Kelaniya, Kelaniya GQ 11600, Sri Lanka
e-mail: zoousa@kln.ac.lk

S.S. De Silva and F.B. Davy (eds.), *Success Stories in Asian Aquaculture*,

6.1 Introduction

Sri Lanka (5°55′–9°55′N; 79°42′–81°52′E) is a continental island of about 65,621 km^2 in monsoonal Asia with rich water resources. There are 103 perennial rivers in Sri Lanka, most of which run radially from the central highland area and drain into the western, southern, and eastern coasts. These river basins drain over 90% of the land area. Floodplains, lagoons, and estuaries also form natural inland water resources of the island (NSF 2000). There are no natural lakes in Sri Lanka, but the country has a multitude of reservoirs, including ancient irrigation reservoirs and recently constructed multipurpose reservoirs (De Silva 1988; Fernando 1993). Almost all of these reservoirs are situated in the zone of the country receiving less than 187 cm annual precipitation (Fig. 6.1). This zone is located within the lowest peneplain of the island, and covers approximately 66% of the total land area of the country, which holds 33% of the country's population. Over 90% of the rural population is directly involved in farming and the peasant community is characterized by acute poverty and malnutrition (De Alwis 1983).

The inland fishery of Sri Lanka has been essentially a capture fishery from perennial reservoirs (major and medium-scale irrigation reservoirs; Table 6.1). However, most small-sized (<100 ha) minor irrigation reservoirs in the country dry up for 2–3 months of the year, but can be utilized for the development of culture-based fisheries (CBF). This is essentially a fisheries enhancement strategy (Lorenzen et al. 2001; De Silva 2003) through the stocking of individuals, a practice frequently used by fisheries owners, managers, and scientists throughout the world (Cowx 1994). Although fishery enhancement has received much attention, CBF have not received its due attention, in spite of its potential to contribute significantly to fish production and also in view of its potential impacts on the rural poor in most of the developing nations (De Silva 2003).

CBF are fisheries based mainly or entirely on the recapture of farm-produced seed stocked in water bodies after an adequate growth period (De Silva 2003). CBF combines elements of aquaculture and capture fisheries, and relies entirely on the natural productivity of the water body for growth of the stocked fish, and on artificial stocking as a means of recruitment (Lorenzen 1995).

Fish yield of small reservoirs, where the management is on the basis of CBF, is dependent on a number of parameters dependent on the natural conditions of the water body, such as growth and mortality rates, in addition to the quality of stocking material. A key technological management problem is therefore to identify stocking and harvesting regimes that make the optimal use of the given conditions (Lorenzen 2001). Accordingly, stocking density (SD), size at stocking, size at harvesting, fishing mortality, and harvesting schedule hold the key for optimizing yields (Sugunan 2001; De Silva 2003). CBF is considered to have the potential to contribute significantly and increasingly to fish production and food security in developing countries, and is widely practiced in Asia and other developing countries for food-fish production (De Silva and Amarasinghe 1996; Welcomme 1996; Cowx 1998; Lorenzen et al. 1998; Welcomme and Bartely 1998; Nguyen et al. 2001; De Silva 2003).

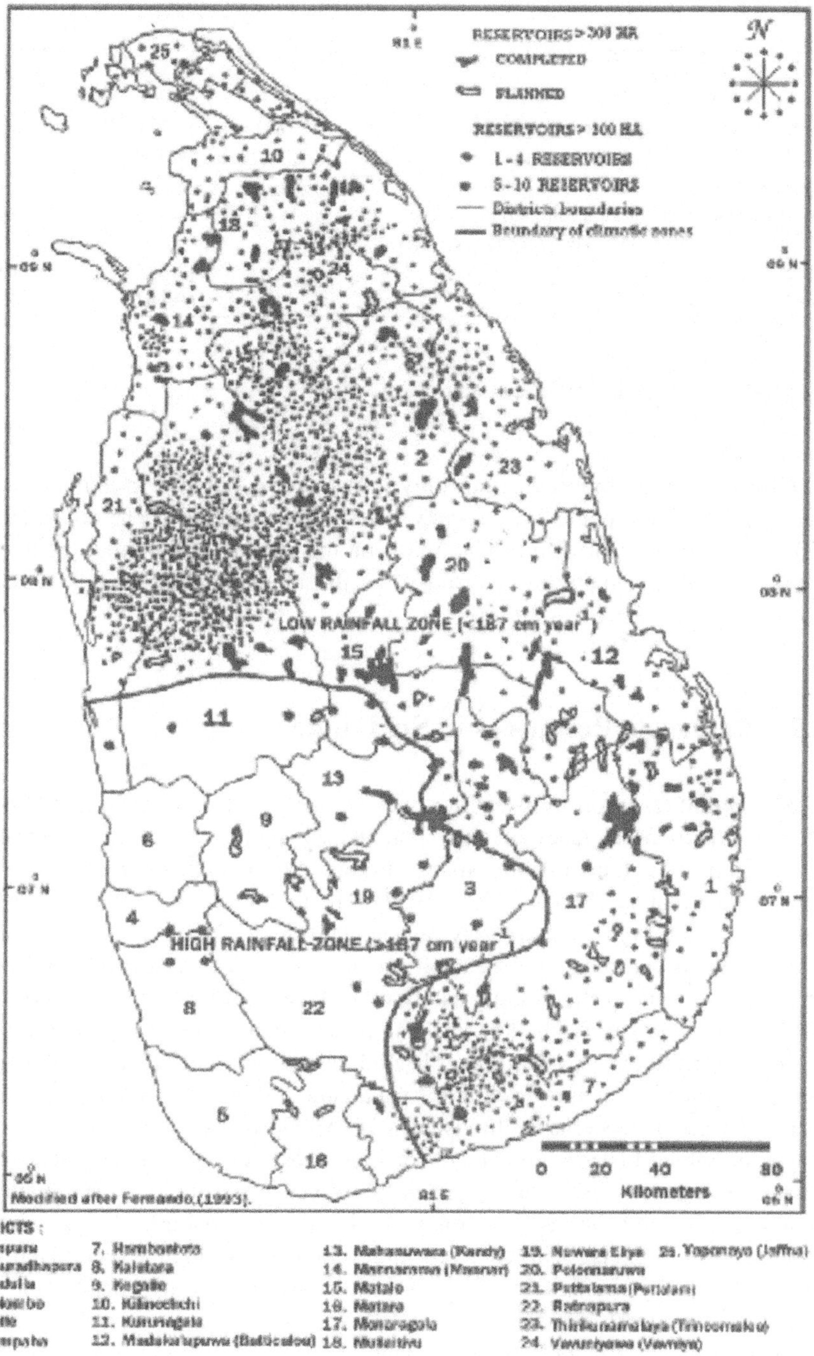

Fig. 6.1 The distribution of reservoirs in the different administrative districts in Sri Lanka, with the line demarcating the dry and wet zones

Table 6.1 The estimated surface area of lentic water bodies of Sri Lanka

Type	Number	Area (ha)	Percent
Major irrigation reservoirs[a]	72	70,850	41.7
Medium-scale reservoirs[a]	160	17,001	10.0
Minor irrigation reservoirs[a]	> 12,000	39,271	23.1
Floodplain lakes (natural)		4,049	2.4
Hydroelectric reservoirs[b]	7	8,097	4.7
Multipurpose reservoirs impounded under the Mahaweli River Basin development project		13,650	8.0
Other		17,023	10.0
Total		169,941	100.0

[a]Ancient
[b]Recent
Modified after Costa and De Silva (1995)

CBF is gaining importance not only in reservoirs, but also in temporary water bodies in the flood plains and ox-bow lakes (De Silva 2003). The contribution of CBF to the total inland fish production is on the increase throughout the tropics; a tendency that is sometimes referred to as the intensification of inland fisheries (Lorenzen 1993; De Silva 2003).

6.2 Reservoir Resources in Sri Lanka

In Sri Lanka, reservoir construction was an integral part of its over 2,500-year-old civilization. Many landmarks still remain, proving the sovereignty of a hydraulic civilization of Sri Lanka, which spans through more than 2,000 years of the country's written history (Brohier 1934, 1937). The hydraulic civilization in Sri Lanka has possibly evolved from an early rain-fed shifting agriculture to small-scale irrigation that has in turn led to the establishment of major irrigation systems. The total area of reservoirs in the country is about 170,000 ha (Costa and De Silva 1995; Table 6.1), averaging 2.7 ha km^{-2} of land area, which is perhaps the highest in the world (Fernando 1993). The villages in Sri Lanka have traditionally developed various management practices leading to sustainable utilization of fishery resources in village irrigation systems (Ulluwishewa 1995). However, there were no commercial-scale inland fisheries in ancient Sri Lanka, as there were no sizeable populations of indigenous fish recruitment in the reservoirs of the country which could support profitable fisheries. The commercial scale, artisanal inland fishery in larger perennial reservoirs is relatively a recent development during the second half of the twentieth century that is considered to have resulted with the introduction of the cichlid, *Oreochromis mossambicus,* in 1952. This exotic species had significantly contributed to the dramatic increase of inland fish production in Sri Lanka, as it accounted for over 80% of the landings in the reservoir fisheries (Fernando and Indrasena 1969; De Silva 1988).

6.3 Water Resources for Culture-Based Fisheries Development in Sri Lanka

An estimated 12,000 village reservoirs, of which 10,000 are functional for irrigating agricultural lands, are found in Sri Lanka (Anon 2000). These small reservoirs, in addition to their primary uses, i.e., irrigation, are used for various communal activities such as bathing, drinking-water supply, and buffalo and cattle keeping. CBF development in these water bodies should therefore be carried out without impeding their multiple uses.

Small-scale water conservation systems, referred to as village/small reservoirs, created by constructing earthen bunds across natural drainage basins, are a distinctive feature of the low-rainfall zone of Sri Lanka. These reservoirs depend entirely on direct rainfall and runoff water from their own catchments. These reservoirs, total an acreage of 39,300 ha (Mendis 1977, Table 6.1). These occur in the form of distinct cascades that are positioned as either well-defined small cascades or meso-catchment basins (Udawattage 1985; Panabokke 2001). A cascade is defined as a connected series of tanks organized within the meso-catchments of the dry-zone landscape, storing, conveying, and utilizing water from an ephemeral rivulet (Madduma Bandara 1985). These reservoirs irrigate paddy fields along the channels (Panabokke et al. 2002; Fig. 6.2). A great majority of these small village reservoirs are less than 100 ha in surface area at full supply level (FSL), and are distributed across the undulating landscape of the low rainfall zone.

The main source of livelihood of the majority of the people in the low-rainfall zone is cultivation of crops (mainly paddy) and allied activities in the two agricultural farming seasons ("*Yala*" from March to June and "*Maha*" from October to February), based on the annual rainfall pattern. Hence, the majority of the people in the low-rainfall zone have two peak working seasons in a year. Between those two labor-intensive peak seasons, the farmers have sufficient time to be mobilized to CBF activities (De Alwis 1983; Fernando and Halwart 2000; Murray et al. 2001).

6.4 CBF Development in Reservoirs of Sri Lanka

Most of the minor irrigation reservoirs in the low-rainfall zone of Sri Lanka are dry for 2–3 months in August–October every year, and are therefore termed as seasonal reservoirs. Based on a survey carried out in 1962, Mendis (1965) suggested that the small, village reservoirs (minor irrigation reservoirs) are biologically productive, and could be used for enhancing fish production. Subsequently, in 1963, eight reservoirs in Polonnaruwa administrative district were stocked with juveniles of *Chanos chanos* and *O. mossambicus* (Anon 1964). Indrasena (1964) also suggested that these reservoirs can and should be utilized for the development of CBF. Indrasena (1965) reported that high yields were obtained from several seasonal

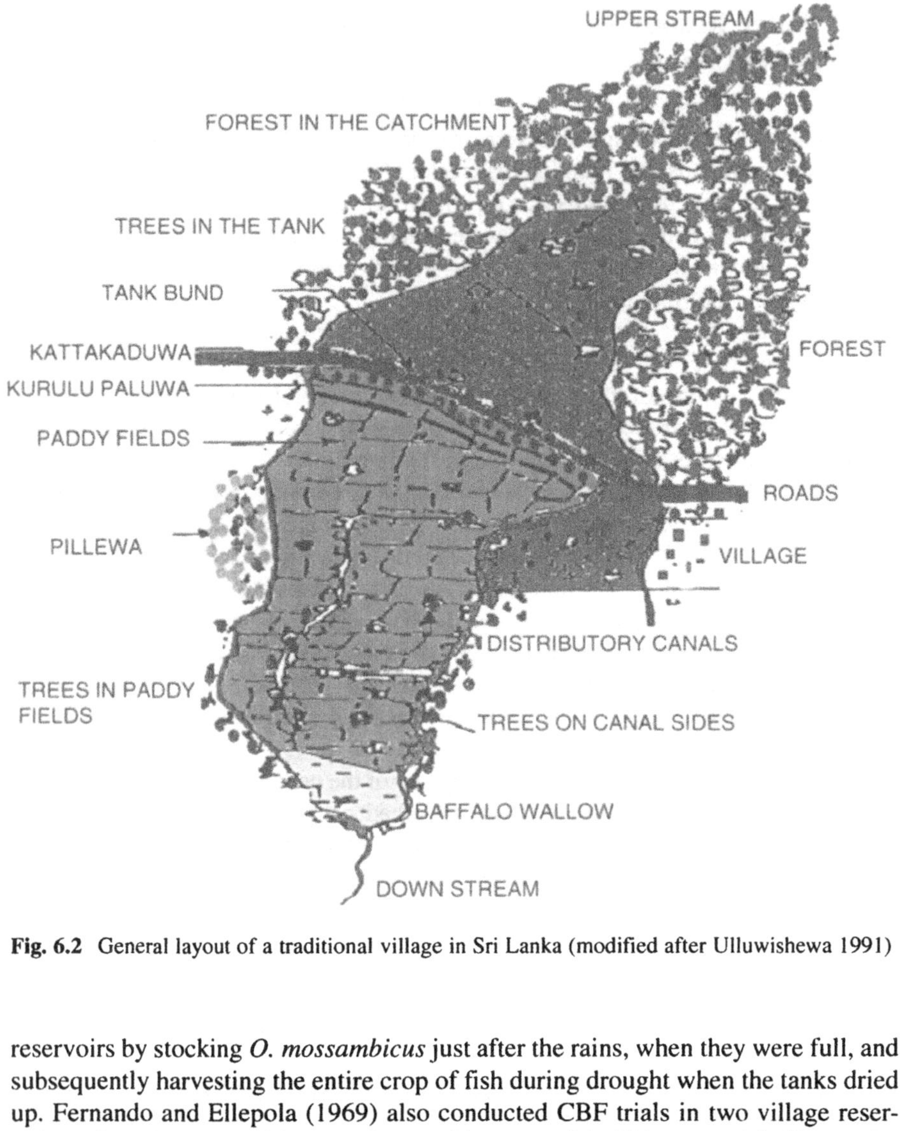

Fig. 6.2 General layout of a traditional village in Sri Lanka (modified after Ulluwishewa 1991)

reservoirs by stocking *O. mossambicus* just after the rains, when they were full, and subsequently harvesting the entire crop of fish during drought when the tanks dried up. Fernando and Ellepola (1969) also conducted CBF trials in two village reservoirs using *O. mossambicus* as the stocking material. Chinese and Indian major carps were not available for stocking village reservoirs at that time, as carp hatcheries were not established in Sri Lanka in the mid-1970s. Mendis (1977) recommended developing CBF in medium-sized and minor irrigation reservoirs as well as seasonal tanks, to develop inland fish production in the country.

The use of seasonal reservoirs for CBF, as highlighted by Indrasena (1964, 1965) and Mendis (1977), was echoed in the late 1970s and early 1980s (Rosenthal 1979; Oglesby 1981), when the government subsequently recognized this strategy as an important approach for enhancing fish production. The initial funding by FAO/UNDO in 1980, (Chakrabarty and Samaranayake 1983), and the subsequent financing by the Asian Development Bank (ADB) (Thayaparan 1982), facilitated the development of CBF in the seasonal reservoirs in the 1980s on a trial basis. In such trials, Chandrasoma and Kumarasiri (1986) reported yields ranging from 220 to 2,300 kg ha^{-1} in 15 seasonal reservoirs (mean 892 kg ha^{-1}) within a growing season.

CBF in the seasonal reservoirs in Sri Lanka only involves stocking of fingerlings after the inter-monsoonal rainy season from December to January, and harvesting of stocked fish during the dry season (August–September). One of the essential components required for the sustainability of CBF in seasonal reservoirs is the availability of seed stock, i.e., fingerlings at the correct time and of the correct size. In the 1980s, all the fingerlings to stock the seasonal reservoirs were produced in the state-owned fish breeding centers.

6.4.1 Why Initial Attempts at CBF Development Failed?

The CBF activities in Sri Lanka in the 1980s were unsustainable in spite of the priorities placed on CBF in the national fisheries development plans. De Silva (1988, 2003) discussed the likely reasons for the overall failure of the strategy, in detail. In addition to the biological productivity of reservoirs and the lack of a guaranteed fingerling supply, the unavailability of appropriate criteria for selecting suitable water bodies was considered a major constraint (De Silva 2003). A politically inspired withdrawal of state patronage for the development of the inland fisheries sector from 1990 to 1994 caused a further major setback to the reservoir fisheries with its centralized management mechanism (De Silva 1991; Amarasinghe 1998). The general collapse of CBF development activities in village reservoirs after discontinuation of government support in 1990 is a clear indication that high dependence on state subsidies for fingerling supply made CBF unsustainable.

In Sri Lanka, induced breeding of major carps is carried out at the state-owned Aquaculture Development Centers (ADCs). The ADCs with carp hatcheries were established in the mid-1970s in Sri Lanka. Possibly owing to the high capital costs involved, there has not been any private-sector involvement in establishing hatcheries for exotic carps. Hence, the government still has a vital role to play in CBF development.

6.4.2 New Thoughts from the Lessons Learnt

Although the government-owned fish breeding centers were leased out to the private sector in 1990, none of the private organizations continued induced breeding of exotic carps, the essential seed stocks for CBF and pond culture. In the fish breeding centers that were leased out to private sector, more lucrative ornamental fish species were reared for export purposes. Amarasinghe (1992) suggested introducing a buy-back scheme for fingerling rearing through community participation.

Table 6.2 Total pond area in ADCs of NAQDA

Location	Mud (m^2)	Cement (m^2)
Udawalawe	82,795	1,998
Dambulla	30,000	3,352
Inginiyagala	35,540	552
Nuwara Eliya	4,570	240
Rambodagalle	7,413	700
Total	160,018	6,842

Source: http://www.naqda. gov.lk

After a revival of state patronage for inland fisheries and aquaculture development in 1994, rearing of fry to fingerling stage for stocking inland water bodies through community participation was also recognized as a feasible strategy, instead of the usual practice of fingerling production in government hatcheries. Induced breeding and rearing of post-larvae upto fry stage (2–3 cm in size) were recognized as the purview of government hatcheries.

To fully revive the state patronage for inland fisheries and aquaculture development, the National Aquaculture Development Authority of Sri Lanka (NAQDA) was established under the Parliamentary Act No. 53 of 1998 and amendment act No. 145 of 2006. Presently, NAQDA is responsible for the development of inland fisheries and aquaculture in the country. Establishment of a strong extension mechanism in NAQDA is one of the major milestones in the process of revival of state patronage for inland fisheries and aquaculture development in the country.

As mentioned earlier, the success of CBF in the seasonal reservoirs depends on the timely availability of quality fingerlings. Induced spawning of major carps is done at the ADCs of NAQDA. Although cement ponds are sufficient to rear post-larvae of major carps up to the fry stage, the mud-pond space available in ADCs (160,018 m^2, Table 6.2) is insufficient to produce adequate quantities of suitably sized fingerlings to stock inland reservoirs. At the full-scale operation of these ADCs, a maximum of about 10 million fingerlings can be produced.

However, active involvement of district and regional aquaculture extension officers (AEOs) of NAQDA, training rural community groups on fingerling rearing in net cages and earthen ponds, considerably facilitated fulfilling fingerling requirements for stocking inland reservoirs. As a result, involvement of the farmers and fishers in rearing of post-larvae of major carps to fingerling size to meet the demand for fingerlings as stocking materials in inland reservoirs has now become a popular income-earning activity among the rural communities.

6.4.3 Legal Framework for Reservoir Management

In Sri Lanka, village reservoirs come under the jurisdiction of the Agrarian Development Act of 2000. A village reservoir is defined as a reservoir whose command area (i.e., irrigable area) is less than 80 ha. These reservoirs are managed by the rural institutions called farmer organization (FO), established under the

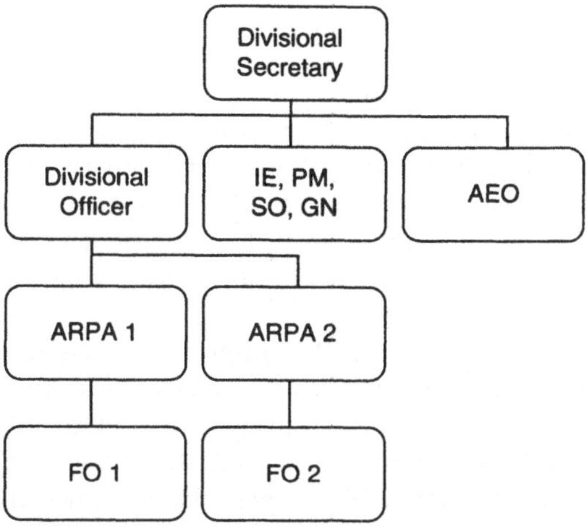

Fig. 6.3 The structure of Divisional Agriculture Committee (DvAC). *AEO* aquaculture extension officer, *GN* Grama Niladhari (village-level administrative officer), *SO* samurdhi officer (government officer responsible for implementing poverty alleviation project), *IE* irrigation engineer, *PM* project manager, *ARPA* agrarian research and production assistant, *FO* farmers organization

above-mentioned Agrarian Development Act. The residents of the village who are involved in agriculture or agriculture-related activities are entitled for membership of the FOs. The divisional officers (DO) of the Department of Agrarian Development (DAD) coordinate the FOs in each village with the help of Agricultural Research and Production Assistants (ARPAs). In most of the districts in the low-rainfall zone of the country, water uses in village reservoirs are highly contentious, because irrigation of rice and other crops is a major activity of FOs. Other economic activities such as CBF development in village reservoirs, therefore, have to be carried out within the constraints of multiple uses of water resources.

Monthly meetings of Divisional Agriculture Committee (DvAC) are held, which are presided by the divisional secretary (DS). DOs, local technical officers, and the office bearers of the FOs also attend these meetings, so that there is a grassroot level involvement in making decisions on the management of village reservoirs. There are legal provisions for various rural development activities through FOs, under the Agrarian Development Act, which includes provisions for the development of CBF in seasonal reservoirs. AEO of NAQDA is also invited to attend the monthly meetings of DvAC. The composition of DvAC is shown in Fig. 6.3.

6.5 Lessons Learnt

CBF development activities that were carried out by the Inland Fisheries Division of the Ministry of Fisheries in the 1980s came to a standstill after discontinuation of the state patronage for inland fisheries development in Sri Lanka in 1990. In the absence of monitoring by the centralized management unit (i.e., Inland Fisheries Division of the Ministry of Fisheries) and the lack of a subsidized fingerling supply for stocking, CBF was not sustained.

Under a project funded by the Australian Centre for International Agricultural Research (ACIAR), a research team from the University of Kelaniya and NAQDA of Sri Lanka, in collaboration with Deakin University, Australia, carried out an extensive study in 45 village reservoirs, focusing on developing holistic management strategies for CBF in village reservoirs incorporating biological, physical, and socioeconomic factors.

Carlson's trophic state index (TSI) (Carlson 1977) measured on the basis of chlorophyll *a* [TSI (Chl-*a*)], was used to classify the village reservoirs based on the biological productivity. As TSI (Chl-*a*) is positively correlated to CBF yield, it is useful for planning CBF development strategies in village reservoirs of Sri Lanka (Jayasinghe et al. 2005). Reservoirs with similar trophic characteristics also showed significant correlation ($p<0.05$) between shoreline/area ratio and CBF yield (Jayasinghe et al. 2006). Also, Jayasinghe and Amarasinghe (2007) demonstrated a positive influence of buffalo density in the vicinity of reservoir on CBF yield. This was owing to the addition of nutrients through cow dung into the village reservoirs.

Naturally recruiting fish species from the associated inlet waters, especially in those village reservoirs that do not dry completely, also influence the performance of stocked species in CBF. In reservoirs where naturally recruited predatory fish species such as *Ophicephalus striatus* and *Mystus keletius* occur in sufficient numbers, the CBF harvests of stocked species are low owing to their high mortalities (Wijenayake et al. 2005, 2007). As density-dependent factors also influence the CBF yields, the optimal SD is found to be about 3,500 fingerlings per ha (Wijenayake et al. 2005, 2007). It should be noted that the optimal SD in CBF is also dependent on the biological productivity of the reservoirs.

Being a communal activity, socioeconomic factors are also equally important for the development of CBF in village reservoirs. It has been found that small group size and homogeneity with regard to caste, kinship, and political ideology of the farming communities, and the level of education and good leadership qualities of the group members have positive influences on the attitudes of rural communities towards the uptake and development of CBF in village reservoirs (Kularatne et al. 2008).

6.5.1 Marketing and Economics of CBF

The *Kanna* meeting (a community meeting held at the beginning of the cropping season) of the FOs is a major event in the village. At this meeting, planning of agricultural activities takes place, and collective decisions are made which cannot be changed by the individuals until the end of the cultivation season, except under special circumstances. With the new provisions available for the development of CBF in the Agrarian Development Act of 2000, important decisions on CBF activities are also made collectively. In most instances, aquaculture management committees (AMCs) are established among the FOs, and strategies for stocking, guarding, and harvesting are decided. The members arrive at agreements on sharing of CBF profit between the fish farmers and agricultural farmers. Levy paid by AMC (generally, about 5% of the profit) to FOs is often used for maintenance of the village reservoir.

The costs involved in CBF and financial benefits derived were evaluated in 23 village reservoirs, where reliable information on stocking, total harvest, and market values of fish could be gathered (Mr. M.G. Kularatne, pers. comm.). The cost of fingerlings was estimated from the unit price of Rs. 1.50 per fingerling (Rs. 100 = 1 US$), which has been fixed by NAQDA (present price of a fingerling is Rs. 2.00). Other costs include that of packing, transport, aquaculture license fee, cost of guarding stocked fish, and hiring seine nets for harvesting. Also, the levy (generally, about 5% of profit) paid to the FO of the reservoir by the AMC was also considered as part of the production cost, because this amount is paid by AMC to FO from their earnings. The group size of AMC varies between 5 and 25, but mostly less than ten members. The net profit (= value of harvest – total cost) is shared equally among the members of the AMC.

The CBF harvest varied widely in the 23 village reservoirs (Table 6.3). It should be noted that some of the reservoirs are connected to seasonal or perennial streams, so that natural recruitment of various fish species also occurs. The fish harvests in some reservoirs therefore consisted of stocked fish and naturally recruited fish. As in many reservoirs, 10% of the stocked fish was *O. niloticus*, and their offsprings also contributed to the final CBF harvest. High yields in reservoirs where less than 4,000 fingerlings were stocked (Table 6.3) might be due to the contribution of non-stocked fish and reproduction of stocked tilapias, to the CBF harvest. The price of fish at the harvesting site ranged from Rs. 30 to Rs. 75 per kg. From every harvest, villagers were provided with fish for home consumption, free-of-charge, and this portion of the harvest were significant and ranged from 3 to 47% of the total. The net profit (both for commercial harvest and total harvest), except in two village reservoirs, where there was a financial loss owing to low survival and recovery rates of the stocked fish, ranged from Rs. 47,372 to Rs. 729,339 (Rs. 100 = US$ 1). The net profit in 17 reservoirs was above Rs. 100,000 (Table 6.3). The financial loss in the two reservoirs was solely owing to the high mortality of the stocked fish as a result of heavy drawdown of the reservoir water level. The most common price of fish at harvesting point was Rs. 30–40 per kg. In a few cases, the prices were higher, up to Rs. 70, especially where the catches were low (Fig. 6.4), but the relationship between the farm-gate price and harvest levels was not significant ($p > 0.05$). Although there were price differences between the villages, the price did not appear to influence the choice of reservoirs for CBF. The income generated from CBF is essentially dependent on the biological productivity of each reservoir and the development of good management practices to enhance recovery rates of the stocked fish.

On an average, 86% of the CBF harvest is sold at distant markets. An average proportion of about 13% of the harvest is taken by villagers during harvesting. The mobile retailers also sell about 1% of the total harvest within the village (Fig. 6.5).

Table 6.3 Costs and net income of CBF in 23 village reservoirs in 2003–2004 culture period

District/Reservoir (ha)	No. stocked	Harvest [commercial] (kg)	Harvest [subsistence] (kg)	Total harvest (kg)	Cost of fingerlings (Rs.)	Other operational costs (Rs.)	Total cost (Rs.)	Value of commercial harvest (Rs.)	Value of subsistence harvest (Rs.)	Total value of harvest (Rs.)	Net total income (commercial + subsistence) (Rs.)	Total commercial net income (Rs.)
Anuradhapura												
Bulankulama (10.4)	12,345	3,671	1,830	5,501	18,517	11,000	29,517	146,848	73,200	220,048	190,531	117,331
Gambirisgaswewa (16.6)	11,520	2,108	200	2,308	17,281	1,500	18,781	84,328	8,000	92,328	73,547	65,547
Karabegama (9.3)	36,317	493	70	563	54,475	2,500	56,975	19,716	2,800	22,516	−34,459	−37,259
Katugampalagama (16.7)	11,740	4,058	500	4,558	17,610	1,000	18,610	162,324	20,000	182,324	163,714	143,714
Pahalasandana-nkulama (21.3)	41,194	4,154	1,000	5,154	61,791	6,000	67,791	145,373	35,000	180,373	11,2581	77,581
Hambantota												
Gonnoruwa (22.4)	63,728	1,232	80	1,312	95,592	16,140	111,732	80,080	5,200	85,280	−26,452	−31,652
Kudaindiwewa (13.4)	58,585	5,052	500	5,552	87,877	1,300	89,177	328,367	32,500	360,867	271,690	239,190
Lunuweraniya (10.8)	46,451	7,679	800	8,479	69,676	1,500	71,176	383,940	40,000	423,940	352,764	312,764
Medagankadawara (16.7)	8,317	1,202	150	1,352	12,475	5,900	18,375	90,180	11,250	101,430	83,055	71,805
Palujadura (4.9)	2,406	1,352	1,200	2,552	3,609	12,000	15,609	67,620	60,000	127,620	112,011	5,2011
Svodagama (3.2)	701	5,763	400	6,163	1,051	300	1,351	172,896	12,000	184,896	183,545	171,545
Weliwewa (5.7)	3,876	2,126	358	2,484	5,814	2,000	7,814	74,414	12,530	86,944	79,130	66,600
Wawegama (18.9)	25,799	3,440	300	3,740	38,698	400	39,098	137,592	12,000	149,592	110,494	98,494
Kurunegala												
Kumbalporuwa (9.5)	2,328	1,948	1,500	3,448	3,491	5,000	8,491	68,163	52,500	120,663	112,171	59,671
Kekunawa (10.0)	19,360	6,430	1,440	7,870	29,040	2,000	31,040	225,050	50,400	275,450	244,410	194,010

In 2004, US$ 1 ≈ Sri Lankan Rupees 100

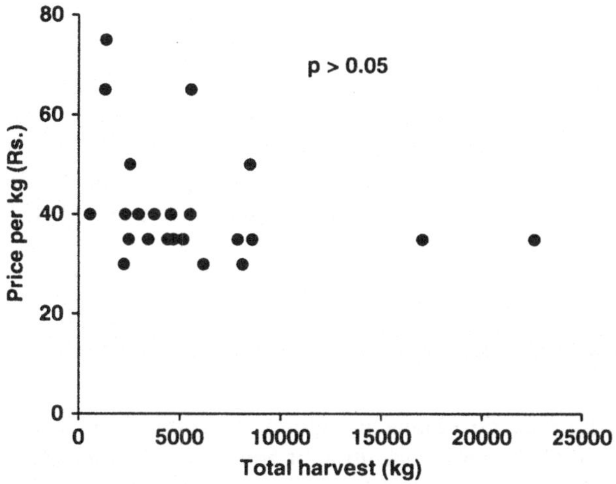

Fig. 6.4 Scatter diagram showing the relationship between fish price and harvests as the sites (Mr. M.G. Kularatne, pers. comm.).

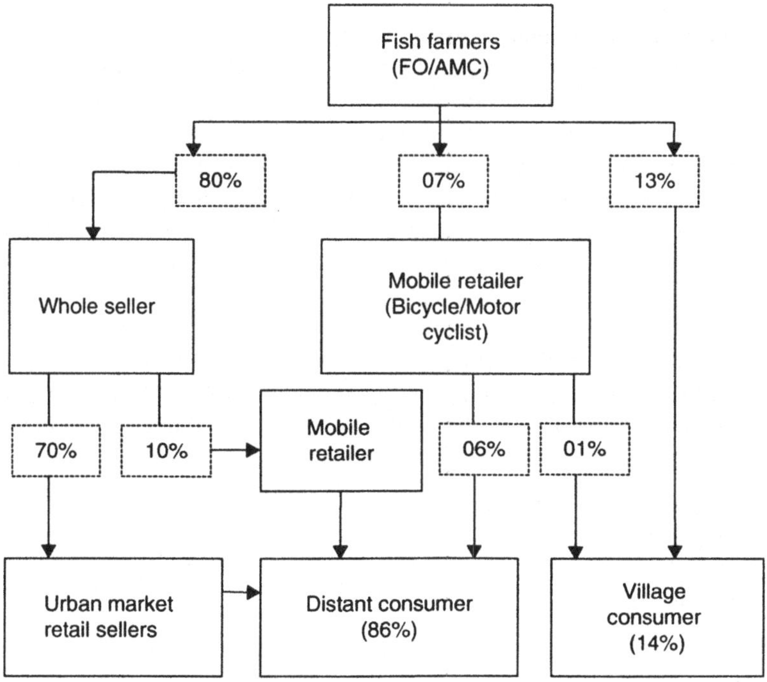

Fig. 6.5 Marketing channels for CBF produce in Sri Lanka (Mr. M.G. Kularatne, pers. comm.)

As the wholesalers buy the major proportion of the harvest (86%) packed in ice to transport to distant markets in lorries, and at the present pace of development of CBF, marketing has not been a problem. However, at the full-scale development of CBF, there is a possibility that markets could be flooded (De Silva 1988). Hence, staggered harvesting, extended over 2–3 months toward the end of the culture period before the reservoir gets dry, to prevent gluts and development of post-harvest technology for value addition and increasing shelf-life, should be considered.

6.5.2 Dissemination of Major Research Findings

The strategy for CBF development requires selection of suitable reservoirs for the development of CBF and organization of aquaculture committees within the rural communities, in consultation with farmer communities. The outcome of the ACIAR-funded project included the following:

- Determination of effective area of seasonal tanks for the organising of CBF on the basis of information published in the data books for the village irrigation schemes of Sri Lanka, ministry of agriculture and land, Department of Agrarian Development.
- Determination of optimal stocking densities and species combinations. The optimal SD for the development of CBF in seasonal reservoirs was determined from a regression relationship (i.e., a second-order curve) between the CBF yield and SD, assuming that SD has the major influence on the CBF harvest. Accordingly, optimal SD was found to be 3,500 fingerlings per ha. Here, the area of the seasonal reservoirs should be the effective area (i.e., 50% of the area at FSL). The reported reservoir area (Anon 2000), which is underestimated and is about half of the area at FSL, can therefore be taken to be approximately as the effective area for CBF. Although it is a fact that the species combination for stocking has to be determined inter alia on the basis of the biological productivity and availability of different habitat types for various fish species, for extension purposes at the field level, general guidelines are needed on the appropriate species combination. Hence, the following species combination can be adopted for stocking (Wijenayake et al. 2005, 2007):

  - 30% Common carp and mrigal
  - 30% Bighead carp/Catla
  - 30% Rohu
  - 10% Nile tilapia

  GIFT strain of Nile tilapia was not found to be superior to non GIFT strains, and hence, the use of GIFT strain for CBF is not advocated (Wijenayake et al. 2007).

- Criteria for the selection of suitable reservoirs for the development of CBF: The suitability of the seasonal reservoirs for the development of CBF can be assessed

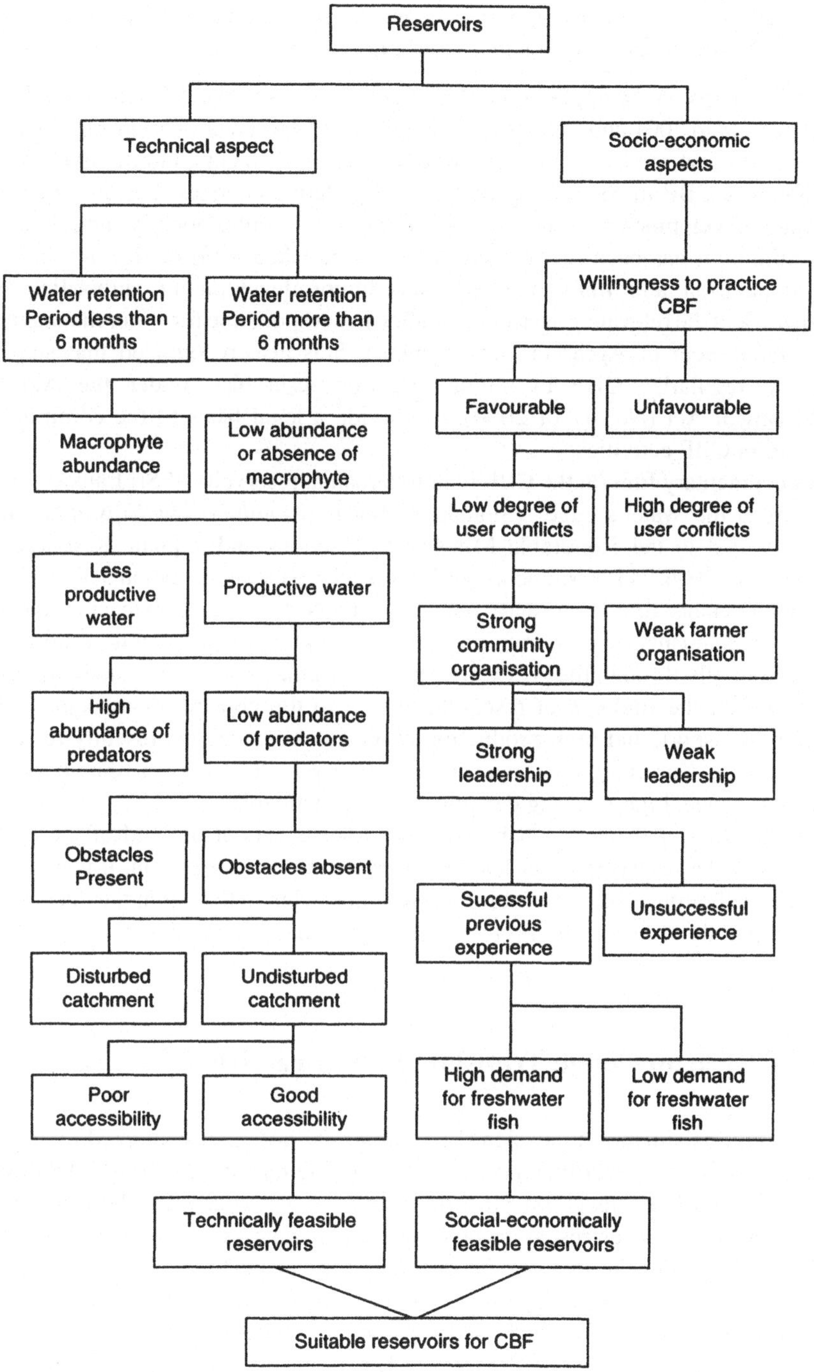

Fig. 6.6 A simplified scheme for selecting suitable village reservoirs for the development of culture-based fisheries (adopted from Jarchau et al. 2008)

on the basis of technical and socioeconomic aspects. Figure 6.6 provides a simplified scheme for selecting suitable village reservoirs for the development of CBF.

Two workshops on the application of research findings of the ACIAR-funded project were held in 2001 and 2003 to bring together all the stakeholder groups involved in CBF. At these workshops, a number of crucial decisions pertaining to the popularization of CBF in Sri Lanka, including legislative changes that are needed to facilitate developments were arrived at and were subsequently implemented. Accordingly, it has been recognized that steps be taken to formalize and improve the cooperation and links between aquaculture development authorities (i.e., NAQDA) and the relevant government authorities responsible for water management (i.e., Department of Agrarian Development). It is also recognized that steps be taken to regularize CBF in minor irrigation reservoirs, under the Agrarian Development Act (No. 47 of 2000), to provide a legal basis to the communities involved in CBF activities.

A documentary film on the CBF in non-perennial reservoirs of Sri Lanka entitled, "Visitors at the Water's edge" (Sinhalese: *Wewata Amuththo Ewith*) was produced and presented to the Honorable Minister of Fisheries and Aquatic Resources on February 10, 2006. This documentary film was telecasted in the national TV channel. This documentary film is presently being used by NAQDA as an extension material.

The mechanisms of disseminating research findings to the rural communities were essentially through the AEOs who provide technical advice on stocking densities based on the findings of research. Owing to the constraints associated with packing, transport, and high production costs, stocking size is presently restricted to about 5 cm. Currently, fish fingerlings are transported in oxygen-filled polythene bags, and hence, high-packing densities are impracticable.

As the stocked fishes are harvested using seine nets at the end of the culture period after the reservoir water level recedes during the dry season, it is necessary to make the bottom of the reservoir free from obstacles to fishing in the deepest part where a puddle of water often remains during the harvesting season.

6.6 CBF Knowledge Translated into Practice

It is a common situation that scientific findings are not readily adopted by the top-level and mid-level government officials for varying reasons. In the ACIAR-funded project for the CBF development, efforts were made to address the issue of the knowledge gap between research and development during the planning phase of the project, as a part of the influencing strategy in the project. As NAQDA is responsible for the CBF development in village reservoirs, the National Project Coordinating Chair was assigned to the Director General of NAQDA.

The ADB funded "Aquatic Resources Development and Quality Improvement Project" commenced in 2002 aimed at the development of inland fisheries in Sri Lanka, and the development of CBF as a major component. The findings of

ACIAR-funded project on fish fingerling rearing in ponds and net cages through community participation, organization of aquaculture committees of FOs through participatory rural appraisal approaches, and introduction of effective extension mechanism for CBF development have therefore been applied directly in the activities of ADB-funded project. The number of net cages or ponds that has to be maintained for stocking a certain extent of reservoirs was determined on the basis of optimal stocking densities, survival rates, and optimal period of culture (Fig. 6.7; Pushpalatha 2006). As different parties are involved in induced breeding, fingerling rearing, and stocking in village reservoirs, linking of these parties is necessary so that effective extension mechanism is the key to the success of the sustainability of CBF in Sri Lanka.

The rural communities engaged in CBF are essentially rice paddy farmers, so that they are not unemployed. However, as mentioned earlier, between the two-labor intensive peak paddy cultivation seasons, the farmers have sufficient time to be mobilized to CBF activities. As such, the CBF has provided them an additional source of income (See Plate 6.1 and Box 6.1). All village reservoirs, which were utilized in the ACIAR-funded study, continue to be engaged in CBF by the respective farming communities, and the project has also stimulated neighboring water bodies to uptake CBF. As the word has spread, other farming communities have realized that the CBF activities are profitable ventures. Consequently, in 2006–2007 culture cycle, 136 rural communities throughout the country have stocked seasonal reservoirs (details in Table 6.4).

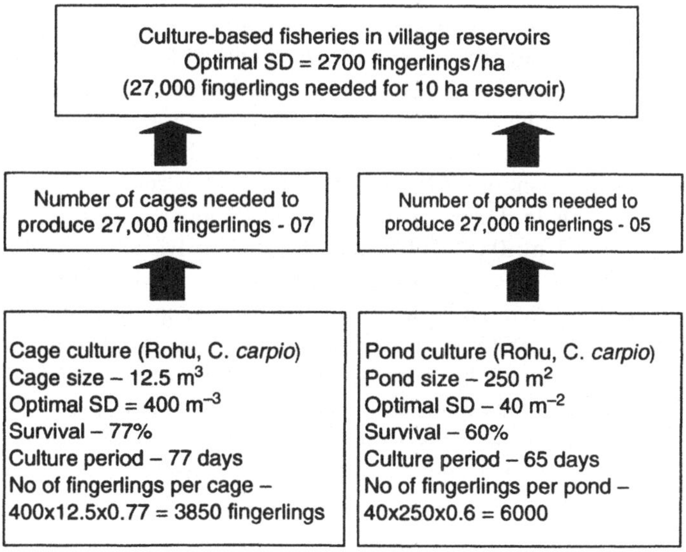

Fig. 6.7 The number of net cages or ponds to be maintained to stock 10 ha of reservoirs. *SD* stocking density (Pushpalatha 2006)

Plate 6.1 CBF harvest in Dozer wewa, Sri Lanka in 2003

Box 6.1 The CBF Development in Dozer wewa (Kukulkatuwa wewa), Sri Lanka

In the late 2002, the research team of the University of Kelaniya and NAQDA visited several reservoirs and participated in the Kanna meetings of the FOs, the community meetings held at the beginning of the cropping season, which is the major event in the village. At these meetings, the ACIAR-funded project on the development of CBF through community participation was introduced to the FOs.In Dozer wewa (or Kukulkatuwa wewa) (13.58 ha; Maximum depth 3 m; 06°26′37.54″N, 81°19′17.34″E), all the agricultural farmers, except one (Mr. Karunadasa) showed their unwillingness to adopt CBF. Mr. Karunadasa, perhaps, owing to his entrepreneurship attitude, showed his willingness to adopt CBF in Dozer wewa. Accordingly, he has stocked the reservoir with the courtesy of the FO of this reservoir in December 2002. The species combination used was 30% common carp, 40% bighead carp, 20% rohu/mrigal, and 10% freshwater prawn. The stocking density was 2,165 fingerlings per ha. The total number of fingerlings stocked was 29,500 and the cost of fingerlings was Rs. 44,250. The ACIAR-funded reservoir fisheries project paid 50% of the fingerling cost. The composition of the harvest (in kg per ha) at the end of the culture period of about 9 months in September 2003 was as follows:

(continued)

Box 6.1 (continued)

Common carp – 328.6
Mrigal – 73.2
Bighead carp – 794.7
Rohu – 317.5
Total – 1514.0

As the effective reservoir area (i.e., 50% of the reservoir area at FSL) is 13.6 ha, the total harvest was about 20,600 kg. With the farm-gate price of Rs. 40 kg^{-1}, the value of this harvest was about Rs. 825,000. The net profit was therefore over Rs. 800,000. He pays 3% of his profit to the FO after every harvest.

This financial boost motivated Mr. Karunadasa to take to CBF, and he continued to stock the reservoir annually. The number of fingerlings stocked in the subsequent years (29,000 in 2003, 20,000 in 2004, 25,000 in 2005, 30,000 in 2006, and 25,800 in 2007) was more or less the same. He has become one of the richest in the village through CBF in Dozer wewa. In 2005, however, this reservoir has now become a perennial reservoir after the diversion of water from an irrigation development project (Galamuna Irrigation Development Project). This has changed the species composition in the reservoir, because carnivorous fish species such as *Glossogobius giuris*, *Ompok bimaculatus* also entered into reservoir. After this water diversion, evidently, the CBF harvest in the reservoir declined from 20,600 kg in 2003 to 8,700 kg in 2004, 9,760 kg in 2005 and 6,400 kg in 2006. Mr. Karunadasa's success in CBF in Dozer wewa was attributed to his own interest and personal capacity, knowledge about local resources, management skills, social integration and harmony, and institutional (NAQDA) support.

In April 2005, Mr. Karunadasa established a mini-nursery to rear fish fry upto fingerling size to stock in Dozer wewa. In his mini-nursery, there is a capacity to rear about 180,000 fingerlings in a culture cycle. Mr. Karunadasa buys fish fry from the state-owned aquaculture development centre in Udawalawe at the cost of 25 cents per fry. Seven members of FO assist him to maintain the mini-nursery. Three culture cycles are carried out per year to rear fry upto fingerlings and on average 100,000 fingerlings are produced per culture cycle. The selling price of the fingerling is Rs. 2.00. Considering the cost of feeding of fingerlings using locally available feed, such as rice bran, the net income from fingerling production can be estimated to be over Rs. 100,000 per culture cycle.

Establishment of community-based mini-nurseries for rearing fish fingerlings is one of the activities of ADB-funded project. As minor-perennial reservoirs (<200 ha) are also stocked as a part of the CBF development strategy of NAQDA and ADB-funded project, the period of fingerling requirement for stocking is not necessarily restricted to post inter-monsoonal period, December–January. Therefore, there is a ready demand for fish fingerlings in most districts and through the effective extension mechanism of NAQDA, selling of fingerlings reared in

Table 6.4 Number of reservoirs and their total extents (50% of the area at full supply level, treated as effective area) and the number of fish fingerlings stocked at the total harvests during 2006–2007 culture cycle in different districts of Sri Lanka

District	Number stocked	Extent stocked (ha)	Fingerlings stocked	Harvest (kg)
Anuradhapura	6	63	145,095	28,399
Badulla	11	35.4	37,000	11,725
Hambantota	5	46.6	25,625	2,505
Kurunegala	48	576	503,515	55,098
Matale	11	70	87,130	19,271
Moneragala	23	134	235,080	67,950
Polonnaruwa	3	27	90,000	2,540
Puttalam	25	314	173,130	32,098
Ratnapura	4	27	28,050	4,238
Total	136	1,323	1,314,625	213,824

Source: Ms. J.M. Asoka (pers. comm.)

mini-nurseries is considerably facilitated. Several mini-nurseries were established in nine administrative districts for fingerling rearing under the auspices of ADB-funded project (Weerakoon 2007). Establishment of these mini-nurseries was facilitated by the ADB-funded project through the extension mechanism of NAQDA (Table 6.5).

Realizing the fact that this strategy is a promising means to improve the livelihoods of rural people. In several other organizations also got involved in the development of CBF. Under the FAO-funded Special Program for Food Security, CBF development activities were carried out in 2002–2006 in the North Central Province of Sri Lanka (http://www.fao.org/TC/spfs/srilanka/intro_en.asp). The GTZ-funded "Fisheries Community Development and Resource Management Project" carried out in the southern province of Sri Lanka in collaboration with NAQDA, investigated various aspects of participatory approaches to CBF development (Wijeyaratne and Amarasinghe 2008). Several NGOs such as Sewa Lanka Foundation have also carried out CBF activities in several provinces of the country (Creech et al. 2002). All these projects facilitated introduction of CBF into rural areas of the country.

Cyprinus carpio was the species stocked in highest number in the 2006–2007 culture cycle (Fig. 6.8). The fingerlings of *C. carpio* are easy to rear, and consequently, most mini-nursery owners prefer to raise this species. Obviously, the major factor that influences the choice of fish species for stocking in Sri Lankan village reservoirs is the fingerling availability. Despite the low numbers stocked, *Catla catla* (catla) registered the highest harvest owing to its faster growth rate. *Labeo rohita* (rohu) is also one of the popular species among the fish farmers of mini-nurseries, and *O. niloticus* and *Aristichthys nobilis* (bighead carp) also resulted in significant harvests. These two species, GIFT strain of *O. niloticus*, *Ctenopharyngodon idella* (grass carp), *Hypophthalmichthys molitrix* (silver carp), and *Labeo dussumieri* were produced in state-owned fish hatcheries.

Table 6.5 Details of the mini-nurseries in operation in 2008

District	Name of the mini-nursery	No. of ponds	Area (m^2)	Construction date	Culture cycles	No. of fry stocked	Fingerlings harvested
Ampara	Raja wewa	13	3,270	12-01-2007	4	1,162,800	579,526
	Kirawana	9	3,300	09-05-2007	2	691,050	468,973
	Padaviya	10	2,250	21-05-2005	6	662,000	276,035
	Mahadiul wewa	09	2,750	12-01-2005	3	654,800	360,957
Badulla	Nawameda-gama	10	3,100	23-06-2005	6	1,491,675	725,854
	Serupitiya	10	4,000	12-07-2006	5	1,167,950	552,946
Hambantota	Andiyagama wewa	08	2,600	30-04-2005	NA	NA	NA
	Divul wewa	10	3,340	26-10-2006	NA	NA	NA
	Ridiyagama	11	3,390	13-12-2005	NA	NA	NA
Kurunegala	Hakwatuna Oya	09	2,680	06-04-2005	9	1,493,340	710,833
	Kathnoruwa	10	3,480	12-09-2005	6	1,281,210	497,288
	Pahala Halmillewa	10	2,880	NA	2	470,000	211,806
Matale	Meewelpitiya	10	2,000	15-02-2007	3	346,000	66,681
Moneragala	Kesellanda	06	1,950	20-07-2004	9	1,062,160	452,550
	Muthukandiya	10	4,000	06-12-2006	3	443,741	227,788
	Dozer wewa	09	2,900	27-04-2005	11	1,259,200	557,399
	Alugalge	10	3,550	27-06-2006	6	825,500	343,740
Polonnaruwa	Elle wewa	10	3,100	14-12-2005	6	1,746,748	602,824
	Sewanapitiya	12	4,200	09-09-2006	3	746,480	363,142
Puttalam	Devala-handiya	10	4,000	04-04-2006	4	1,248,120	592,413
	Wijaya-katupotha	08	3,200	NA	1	142,000	56,615
Trincomalee	Pareipanchan wewa	07	3,772	NA	NA	NA	NA

NA not available

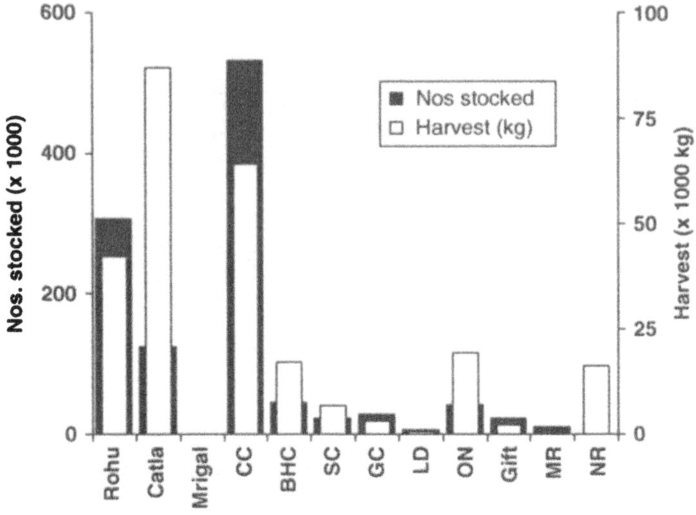

Fig. 6.8 Species-wise breakdown of the number of fingerlings stocked and harvested in village reservoirs of Sri Lanka during 2006–2007 culture cycle. Rohu – *Labeo rohita*, Catla – *Catla catla*, Mrigal – *Cirrhinus mrigala*, CC – *Cyprinus carpio* (Common carp), BHC – *Aristichthys nobilis* (Bighead carp), SC – *Hypophthalmichthys molitrix* (Silver carp), GC – *Ctenopharyngodon idella* (Grass carp), LD – *Labeo dussumieri*, ON – *Oreochromis niloticus*, *Gift* genetically improved farmed tilapia (GIFT) strain of *O. niloticus*, MR – *Macrobrachium rosenbergii*, *NR* naturally recruited species (Source: Ms. J.M. Asoka, pers. comm.)

In recognition of the importance of the ACIAR-funded project's findings with respect to the region, ACIAR has also supported a further project to disseminate these findings through the intergovernmental agency, NACA (Network of Aquaculture Centres in Asia-Pacific), a regional organization comprising 17 member countries. Consequently, three workshops were held in Indonesia, Laos, and Cambodia to disseminate the Sri Lankan and Vietnam project results and to evaluate the adoption of the findings, suitably modified, in other countries. A monograph on better practice approaches for CBF development in Asia (De Silva et al. 2006) was published by NACA (http://library.enaca.org/NACA-Publications/CBF_manual.pdf).

6.7 Vietnam

Recent statistics show that Vietnam has about 4,000 reservoirs with a total area of 340,000 ha (Ngo and Le 2001). Most of these reservoirs were built for the purpose of hydro-electricity generation, flood control, and irrigation. Small reservoirs were mainly built to store stream water to support agricultural activities in dry season.

Fish stocking in reservoirs, of approximately 1,000 ha or less, dates back to 1962 and was carried out through the governmental institutions until the opening of the economy in 1995. However, production of stocked species fluctuated highly, with the contribution to the total production ranging from 15 to 90%.

The Government of Vietnam recognized the importance and potential of reservoir fisheries in meeting the increasing demand for animal protein, as well as providing additional employment and improving the livelihoods of rural communities. Hence, farmers are encouraged to utilize reservoirs for fisheries development, particularly under the new economic regime. As a result, hatcheries were constructed for fingerling production for restocking programs in almost all large reservoirs. Hatcheries were also built for enhancement of smaller reservoirs as well as supplying seed for aquaculture. In addition, small reservoirs were leased to farmers, farmer groups, or local organizations to conduct CBF activities.

CBF, in a very rudimentary form, was practiced in Northern Vietnam, in small reservoirs for a long time. In the case of Vietnam, however, the water bodies were leased to individual farmers for varying periods of time for fishery activities by the Provincial and or District administrations. The lease period ranged widely and there were no strict regulations determining the latter. In such a context, the lessees were rather reluctant to make major investments that would lead to an improvement in the practices. Also, the practices were rather ad hoc, as little was known about the optimal species combinations to be used, the proportions in the species combinations, as well as the other parameters that would influence optimal production.

A 5-year R&D program in small reservoirs in the two provinces, Thai Nguyen and Yen Bai, North Vietnam, enabled fine tuning the scientific approach to stocking and harvesting, as well as caring for the stock. The enthusiasm of the farmers involved in the activity was improved. The dissemination of information on the benefits of adoption of CBF practices through appropriate workshops and print material created interests amongst rural communities in six other provinces.

The major social constraint was recognized as the lack of uniformity in the leasing procedures of the water bodies. With the popularity and the demand for adoption of CBF activities, the provincial and district administrations made appropriate regulatory changes to the lease period, increasing it to a minimum of 5 years. This facilitated and encouraged fish farmers to invest in bringing about required repairs in the reservoirs. For example, fish farmers tend to repair sluices and the like, which all lead to further increases in productivity through the prevention of escape of seed stock. Moreover, CBF was included in the development plans of the Northern Provinces and accordingly, the relevant administrations have taken steps to improve seed stock supplies through improvements to hatcheries. Also, the relevant administrations have taken steps to improve the market chains with a view to ensuring acceptable and profitable farm gate prices. CBF has not only been responsible for an effective secondary use of small water bodies in Northern Vietnam for fish production, but has resulted in generating supplementary income, increasing the food-fish availability at an affordable price to impoverished remote rural communities.

6.8 Lao PDR

The main animal protein source of Laotian people, who have the lowest per capita income amongst Asian nations, is fish. The current fish production, almost entirely based on the very seasonal riverine and reservoir capture fisheries, approximates 30,000 tons per year, and has been static over the past 5 years or so. The Government of Lao PDR is endeavoring to increase the fish consumption to 23 kg $caput^{-1}$ per year by year 2010. The gap between supply and demand should be met from aquaculture-related developments, which is still in its infancy, and currently accounts for less than 25,000 tons per year. In this regard, it has been recognized that substantial increases in fish production could be obtained through effective and optimal utilization of seasonal water bodies, such as flood plain depressions and reservoir coves, for CBF, a practice that requires little or no capital investment, and harnesses natural productivity of these water bodies for augmenting fish production. The practice is also environmentally non-perturbing, when compared with conventional aquaculture.

In the abovementioned context, an R&D program was initiated to develop CBF in small reservoirs, flood plain depressions, and reservoir coves in Lao, with community engagement. The success of the activities in two provinces, Vientiane and Borikamxay, have not only triggered off many communities to adopt CBF, but has spread to many other provinces within a span of 18 months. Most importantly, the Laotian government issued a decree authorizing the use of suitable water bodies for CBF development, at the very beginning, thereby facilitating the communities to take appropriate action to prepare the water bodies for facilitating CBF activities, such as netting the entrance to a reservoir cove and authorizing/legalizing the ownership to the stock.

Adoption of CBF activities in these impoverished communities has increased food-fish availability, provided supplementary income, and most of all, brought about communal harmony and well being. The communities differ in the manner in which each shares the profits gained from CBF. In all the cases, the adoption of CBF has resulted in an increase in fish production from the water bodies by 10–20-folds. However, in almost every instance, a certain percentage of the profit is retained for the purchase of fingerlings in the ensuing year, as well on improving community amenities, such as improving the infrastructure of the village temple, electricity supply, amongst others. Also, each community adopts a chosen strategy for harvesting and marketing the produce, with a view to ensure profits and their equitable distribution.

6.9 Conclusions

CBF essentially is a communal activity, often rural, which utilizes the existing water bodies, for the secondary purpose of enhancing fish production, beyond the levels that are obtained naturally. The secondary use of small water bodies for fish production does not impair and or impede its primary use, which is downstream agriculture. CBF also is an environmentally friendly activity, as apart from stocking, there is little or no manipulation of the environment.

In the wake of increasing constraints on water resources use of existing small water bodies, CBF is gradually accepted as a significant strategy for increasing fish production, particularly benefitting rural communities, to augment their nutrition status, income generation and social well being. The environmentally friendly nature of CBF activities and the minimal capital investments needed are major advantages to further its adoption by rural communities. It is in the above-mentioned context that those developing nations that have begun to adopt this practice as a means of increasing fish production are witnessing an increasing degree of adoption of CBF by rural communities, and needless to say, a gradually evolving success story in aquaculture.

Acknowledgments We acknowledge the cooperation of IISD and NACA for preparing this chapter. Mr. M.G. Kularatne assisted in the preparation of Fig. 6.2 of this chapter and provided unpublished information on socioeconomics of CBF in Sri Lanka. We are thankful to Dr. D.E.M. Weerakoon, Director General of National Aquaculture Development Authority of Sri Lanka (NAQDA), for providing data on the operation of community-based mini-nurseries. Ms. J.M. Asoka provided the CBF data that she has collected for her postgraduate research.

References

Amarasinghe, U.S. 1992. Recent trends in the inland fishery of Sri Lanka, 84–105. In *FAO Fisheries Report No. 458*, ed. Baluyut, E.A. Rome: FAO, 281.

Amarasinghe, U.S. 1998. Reservoir fisheries management in Sri Lanka: achievements, mistakes and lessons for future. *International Review of Hydrobiology* 83(Special Issue): 523–530.

Anon. 1964. *Administrative report of the director of fisheries for 1962–1963,* 84. Colombo: Government Publication Bureau.

Anon. 2000. *Data book for village irrigation schemes of Sri Lanka.* (23 volumes). Colombo: Department of Agrarian Development, Ministry of Agriculture and Lands.

Brohier, R.L. 1934. *Ancient irrigation works in Ceylon*, vol. 1, 37. Colombo: Ceylon Government Press.

Brohier, R.L. 1937. *Ancient irrigation works in Ceylon*, vol. 2, 43. Colombo: Ceylon Government Press.

Carlson, R.E. 1977. A trophic state index for lakes. *Limnology and Oceanography* 22: 361–369.

Chakrabarty, R.D. and R.A.D.B. Samaranayake. 1983. Fish culture in seasonal tanks in Sri Lanka. *Journal of Inland fisheries* 2: 125–140.

Chandrasoma, J.K. and W.S.A.A.L. Kumarasiri. 1986. Observations on polyculture of fish in seasonal tanks in Sri Lanka. *Journal of Inland Fisheries* 3: 49–55.

Costa, H.H., and P.K. de Silva. 1995. Limnological research and training in Sri Lanka: State of the art and future needs. In *Limnology in developing countries*, eds. Gopal, B., and R.G. Wetzel, vol. 1, 63–103. New Delhi: International Association for Limnology Publication, International Scientific Publication.

Cowx, I.G. 1994. Stocking strategies. *Fisheries Management and Ecology* 1: 15–30.

Cowx, I.G.,ed. 1998. *Stocking and introduction of fish*, 456. Oxford: Fishing News Books.

Creech, S., L. Dassanayake, and J. Wijesinghe. 2002. Renowned justification for the promotion of carp aquaculture in seasonal minor irrigation tanks in Sri Lanka. *Sri Lanka Journal of Aquatic Sciences* 7: 23–34.

De Alwis, L. 1983. Possibilities in inland fisheries for developing the peasant economy of the dry Zone. Part II. *Economic Review* 8: 11–12.

De Silva, K.H.G.M. 1991. The predicament of the freshwater fisheries and aquaculture of Sri Lanka. *NAGA, The ICLARM Quarterly* 14: 6–7.

De Silva, S.S. 1988. Reservoirs of Sri Lanka and their fisheries. *FAO Fisheries Technical Paper* 298: 128.

De Silva, S.S. 2003. Culture-based fisheries: an under utilized opportunity in aquaculture development. *Aquaculture* 221: 221–243.

De Silva, S.S. and U.S. Amarasinghe. 1996. Reservoir fisheries in Asia. In *Perspectives in Asian fisheries – A volume to commemorate the 10th Anniversary of the Asian Fisheries Society*, ed. S.S. De Silva, 189–216. Manila: Asian Fisheries Society.

De Silva, S.S., U.S. Amarasinghe, and T.T.T. Nguyen. 2006. *Better-practice approaches for culture-based fisheries development in Asia*, 98. ACIAR Monograph No. 120. Bangkok: Australian Centre for International Agricultural Research, Canberra and Network of Aquaculture Centres in Asia-Pacific.

FAO/UNDP. 1980. *Strategy for development of aquaculture in Sri Lanka*. Report of a planning mission, September–October 1980. ADCP/MR/81/15.

Fernando, C.H. 1993. Impact of Sri Lankan reservoirs, their fisheries, management and conservation. In *Ecology and landscape management in Sri Lanka*, eds. Erdelen, W., C. Preu, N. Ishwaran, and C.M. Madduma Bandara, 351–374. Proceedings of the International and Interdisciplinary Symposium, 12–16 March 1990, Colombo, Sri Lanka. Weikersheim: Margraf Scientific Books.

Fernando, C.H. and W.B. Ellepola. 1969. A preliminary study of two village tanks (reservoirs) in the Polonnaruwa area with biological notes on these reservoirs in Ceylon. *Bulletin of Fisheries Research Station*, Ceylon 20: 3–13.

Fernando, C.H. and M. Halwart. 2000. Possibilities for the integration of fish farming into irrigation systems. *Fisheries Management and Ecology* 7: 45–54.

Fernando, C.H. and H.H.A. Indrasena. 1969. The freshwater fisheries of Ceylon. *Bulletin of Fisheries Research Station*, Ceylon 20: 101–134.

Indrasena, H.H.A. 1964. The development of fresh water fisheries in Ceylon. Proceedings of the symposium on the development of Ceylon's fisheries. *Bulletin of Fisheries Research Station*, Ceylon 17: 287–289.

Indrasena, H.H.A. 1965. Fresh water fishery resources. *Bulletin of Fisheries Research Station*, Ceylon 18: 1–4.

Jayasinghe, U.A.D. and U.S. Amarasinghe. 2007. Buffaloes in favour of culture-based fisheries in Sri Lanka. *Aquaculture Asia* 12(3): 3–7.

Jayasinghe, U.A.D., U.S. Amarasinghe, and S.S. De Silva. 2005. Trophic classification of non-perennial reservoirs utilized for the development of culture-based fisheries, Sri Lanka. *International Reviews of Hydrobiology* 90: 209–222.

Jayasinghe, U.A.D., U.S. Amarasinghe, and S.S. De Silva. 2006. Culture-based fisheries in non-perennial reservoirs of Sri Lanka: Influence of reservoir morphometry and stocking density on yield. *Fisheries Management and Ecology* 13: 157–164.

Jarchau, P., C. Lidzba, and S. Jayasekara. 2008. Feasibility assessment of fish farming in seasonal reservoirs: Packaging technical, social and financial aspects. In *Participatory approaches to reservoir fisheries management: Issues, challenges and policies*, ed. M.J.S. Wijeyaratne and U.S. Amarasinghe, 193–205. Colombo: Sri Lanka Association for Fisheries and Aquatic Resources.

Kularatne, M.G., U.S. Amarasinghe, and S.S. De Silva. 2008. Influence of socioeconomic heterogeneity on culture-based fisheries in non-perennial reservoirs of Sri Lanka. In *Participatory approaches to reservoir fisheries management: Issues, challenges and policies*, ed. M.J.S. Wijeyaratne and U.S. Amarasinghe, 135–150. Colombo: Sri Lanka Association for Fisheries and Aquatic Resources.

Lorenzen, K. 1993. *Assessment and management of South Asian small reservoir fisheries: A population dynamic approach*, 120. London: Marine Resources Assessment Group Ltd.

Lorenzen, K. 1995. Population dynamics and management of culture based fisheries. *Fisheries Management and Ecology* 2: 61–73.

Lorenzen, K. 2001. Using population models to assess culture-based fisheries: A brief review with an application to the analysis of stocking experiments. In *Reservoir and culture-based fisheries: Biology and management*, ed. De Silva S.S., 257–265. ACIAR Proceedings No. 98. Canberra: Australian Centre for International Agricultural Research.

Lorenzen, K., J. Junatana, J. Bundit, and D. Tourongruang. 1998. Assessing culture fisheries practices in small waterbodies: a study of village fisheries in north-east Thailand. *Aquaculture Research* 29: 211–224.

Lorenzen, K., U.S. Amarasinghe, D.M. Bartley, J.D. Bell, M. Bilio, S.S. De Silva, C.J. Garaway, et al. 2001. Strategic review of enhancements and culture-based fisheries. In *Technical proceedings of the international conference on aquaculture in the third millennium*, 20–25 February 2000, eds. Subasinghe, R.P., P.B. Bueno, M.J. Philllips, C. Hough, S.E. McGladdery, and J.R. Arthur, 221–237. Bangkok: Network of Aquaculture Centres in Asia.

Madduma Bandara, C.M. 1985. *Catchments ecosystems and village tank cascades in the dry zone*. Dordrecht: Riedel Publishing Company.

Mendis, A.S. 1965. A preliminary survey of 21 Ceylon lakes. 2. Limnology and fish production potential. *Bulletin of Fisheries Research Station*, Ceylon 16: 7–16.

Mendis, A.S. 1977. The role of man-made lakes in the development of fisheries in Sri Lanka. *Proceedings of Indo-Pacific Fisheries Council* 17(3): 247–254.

Murray, F.J., S. Kodituwakku, and D.C. Little. 2001. Fisheries marketing system in Sri Lanka and their relevance to local reservoir fishery development. In *Reservoir and culture-based fisheries: Biology and management*, ed. De Silva, S.S., 56–65. ACIAR Proceedings No. 98. Canberra: Australia.

Ngo, V.S., and L.T. Le. 2001. Status of reservoir fisheries in Vietnam. In *Reservoir and culture-based fisheries: Biology and management*, ed. De Silva S.S., 29–35. ACIAR Proceedings No. 98. Canberra: Australia

Nguyen, H.S., A.T. Bui, T.L. Le, T.T.T. Nguyen, and S.S. De Silva. 2001. The culture-based fisheries in small, farmer-managed reservoirs in two provinces of northern Vietnam: An evaluation based on three production cycles. *Aquaculture Research* 32: 975–990.

NSF. 2000. *Natural resources of Sri Lanka 2000*. Colombo: National Science Foundation of Sri Lanka.

Oglesby, R.T. 1981. A synthesis of the reservoir fisheries in Sri Lanka. FI: TCP/SRL/8804 Field Document 2. Rome: FAO.

Panabokke, C.R. 2001. The nature and properties of small tank systems of the dry zone and their sustainable production thresholds. In *Food security and small tank systems in Sri Lanka*, ed. Gunasena, H.P.M., 33–47. Proceedings of the workshop organized by the working committee on agricultural science and forestry. September 2000. Colombo: National Science Foundation.

Panabokke, C.R., R. Sakthivadivel, and A.D. Weerasinghe. 2002. Small tanks in Sri Lanka: Evolution, present status and issues, 73. Colombo: International Water Management Institute.

Pushpalatha, K.B.C. 2006. *Strategies for the development of culture-based fisheries in seasonal reservoirs of Anuradhapura district in Sri Lanka, through community participation*, 152. Unpublished M.Phil. Thesis. Sri Lanka: University of Kelaniya.

Rosenthal, H. 1979. Preliminary report and recommendations on reservoir and tank stocking activities in inland fisheries in Sri Lanka. Interim Report to FAO for the Project TCP/SRL/8804, Development of fisheries in the man-made lakes and reservoirs. Rome: FAO.

Sugunan, V.V. 2001. Status of culture-based fisheries in small reservoirs in India. In *Reservoir and culture-based fisheries: Biology and management*, ed. De Silva S.S., 274–280. ACIAR Proceedings No. 98. Canberra, Australia.

Thayaparan, K. 1982. The role of seasonal tanks in the development of freshwater fisheries in Sri Lanka. *Journal of Inland Fisheries* 1: 133–152.

Udawattage, U.D.S. 1985. The development of micro-catchments in Sri Lanka. *Journal of Hydrology* 80: 351–359.

Ulluwishewa, R. 1991. Soil fertility management of paddy fields by traditional farmers in the dry zone of Sri Lanka. *Journal of Sustainable Agriculture* 1(3): 95–106.

Ulluwishewa, R. 1995. Traditional practices of inland fishery resources management in the dry zone of Sri Lanka: Implications for sustainability. *Environmental Conservation* 22: 127–132.

Weerakoon, D.E.M. 2007. Freshwater fish seed resources in Sri Lanka. *FAO Fisheries Technical Paper* 501: 425–440.

Welcomme, R.L. 1996. Stocking as a technique for enhancement of fisheries. *FAO Aquaculture Newsletter* 14: 8–12.

Welcomme, R.L. and D.M. Bartley. 1998. Current approaches to the enhancement of fisheries. *Fisheries Management and Ecology* 5: 351–382.

Wijenayake, W.M.H.K., U.S. Amarasinghe, and S.S. De Silva. 2007. Performance of GIFT strain of *Oreochromis niloticus* in culture-based fisheries in non-perennial reservoirs, Sri Lanka. *Sri Lanka Journal of Aquatic Sciences* 12: 1–18.

Wijenayake, W.M.H.K., U.A.D. Jayasinghe, U.S. Amarasinghe, J.A. Athula, K.B.C. Pushpalatha, and S.S. De Silva. 2005. Culture-based fisheries in non-perennial reservoirs in Sri Lanka: production and relative performance of stocked species. *Fisheries Management and Ecology* 12: 249–258.

Wijeyaratne M.J.S., and U.S. Amarasinghe, eds. 2008. *Participatory approaches to reservoir fisheries management: Issues, challenges and policies*, 235. Colombo: Sri Lanka Association for Fisheries and Aquatic Resources.

Chapter 7
Striped Catfish Aquaculture in Vietnam: A Decade of Unprecedented Development

Nguyen Thanh Phuong and Dang Thi Hoang Oanh

Abstract The aquaculture sector in Vietnam began in the early 1960s with small scale extensive culture systems. The rapid growth of this sector has been achieved during the last two decades as a direct result of its diversification on farming practices and adaptation to the production of exportable species at increased levels of intensification. Foremost amongst these is the catfish culture in the Mekong Delta in Southern part of Vietnam.

The Mekong delta has a total freshwater area of 641,350 ha, which comprises of 67.2% total water surface. This Delta has the most diversified aquatic farming activities and great potential for increasing aquaculture production. There are a few species, which have been commercially produced in the Delta, such as pangasiid catfishes and black tiger shrimp (*Penaeus monodon*). Catfish farming started at the beginning of the 1960s that included Mekong catfish (*Pangasius bocourti*) cultured in small cages and striped catfish (*Pangasianodon hypophthalmus*), locally referred to as basa and tra, respectively, and cultured mostly in small ponds. The actual growth of catfish culture, especially striped catfish, took off at the beginning of 2000 when artificial propagation techniques for striped catfish were developed and mass scale seed production became possible. Striped catfish culture in the Mekong delta is considered a success story of aquaculture in Vietnam, if not globally. The production and export turnover reached 1,200,000 tons worth US $1 billion in 2007. It has triggered the development of a processing sector providing for 150,000 livelihoods, mostly for rural women, and many more in other associated service sectors. The production per unit area averages 400–600 tons/crop, and is probably the highest achieved not only in aquaculture, but in any primary production sector. Moreover, the striped catfish products are exported to over 80 countries world-wide. This success can be attributed to several factors that include: (1) rapid establishment of seed

N. Thanh Phuong (✉) and D.T.H. Oanh
College of Aquaculture and Fisheries, Can Tho University, Campus 2, 3/2 Street, Ninh Kieu District, Can Tho City, Vietnam
e-mail: ntphuong@ctu.edu.vn

S.S. De Silva and F.B. Davy (eds.), *Success Stories in Asian Aquaculture*,

production and culture techniques; (2) establishment and expansion of export markets; (3) development of feed and processing infrastructure;, and (4) supportive policies the government. It is expected that this sector will continue to grow further, albeit at a reduced rate, and will be sustainable.

7.1 Why is it Striped/Tra Catfish (*Pangasianodon hypophthalmus*)?

The Mekong delta in the southern part of Vietnam is known as the region for catfish farming (Plate 7.1). Catfishes in Vietnam comprise two genera, *Pangasius* with ten species and the single species genus *Pangasianodon*. These species also occur in Cambodia, Laos, and Thailand (www.fishbase.org). Of these, *Pangasius bocourti* and *Pangasianodon hypophthalmus* have been farmed in the Mekong delta for decades. The former has been commercially farmed in cages since the beginning of the 1960s, and reached its peak production in 1994 (Phuong 1998). In the early phases of the culture of the catfish species, the seed stock was caught from Cambodian and Vietnamese waters, particularly in the confluence region of the Mekong, Bassac, and Tonle Sap rivers. The second species has been traditionally farmed in small ponds using wild seed for few decades, and commercial culture in cages, pens, and ponds commenced with the development of artificial mass seed production in 2000 (Tuan et al. 2003). These fishes were very common and served as a daily food for many Vietnamese people who live in the southern part of Vietnam, especially in the Mekong Delta.

7.2 Striped Catfish: Life Cycle

Striped catfish (*P. hypophthalmus*) is a migratory species. The fish moves upstream of the Mekong River at the end of flooding season (from October to February) and returns to the main stream at the beginning of the rainy season (from June to August). The spawning ground of striped catfish is generally known to be in the upstream of the Mekong River Delta, more especially, below the Khone Falls on the Lao-Cambodia border (MRC 2002 cited by Trong et al. 2002). The fish spawns at the beginning of the rainy season and the adhesive eggs are deposited on roots of aquatic macrophytes and other types of substrates. The newly hatched larvae drift downstream with the water current and swept into floodplain areas in southern part of Cambodia and the Mekong delta in Vietnam. Striped catfish is an air-breathing species (Browman and Kramer 1985 cited by Cacot 1999), which allows the fish to withstand low levels of dissolved oxygen. However, this fish prefers deep and flowing water, and therefore, farming areas of striped catfish are mostly along the branches of Mekong River banks and large canals (Fig. 7.1).

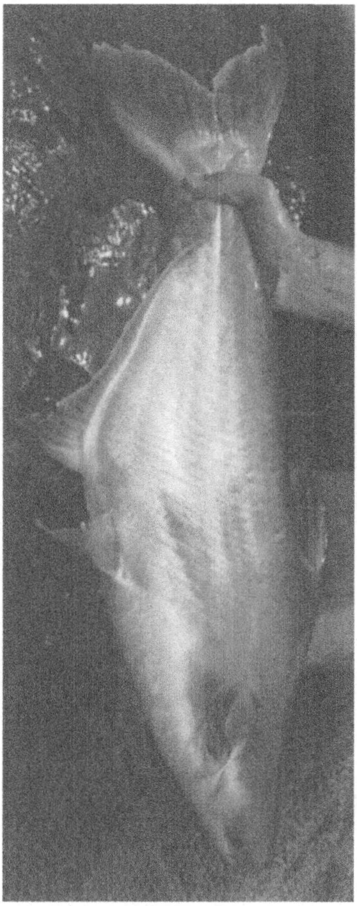

Plate 7.1 Striped catfish (*Pangasianodon hypophthalmus*)

7.3 Striped Catfish: Production Chain

The chain from production to market of striped catfish is completed within the Mekong River delta. The sector can be subdivided into four main components including seed production or hatchery, nursery, grow-out, and processing (Fig. 7.2). These components have changed overtime, but there have been remarkable turning points to create a driving force in the farming of this species. The first turning point is that the Vietnamese and foreign researchers turned their joint efforts to develop artificial propagation techniques of both striped and Mekong river catfishes in early 2000 (Tuan et al. 2003). Lately, the techniques for grow-out and feed and feeding have also been developed from the research achievements.

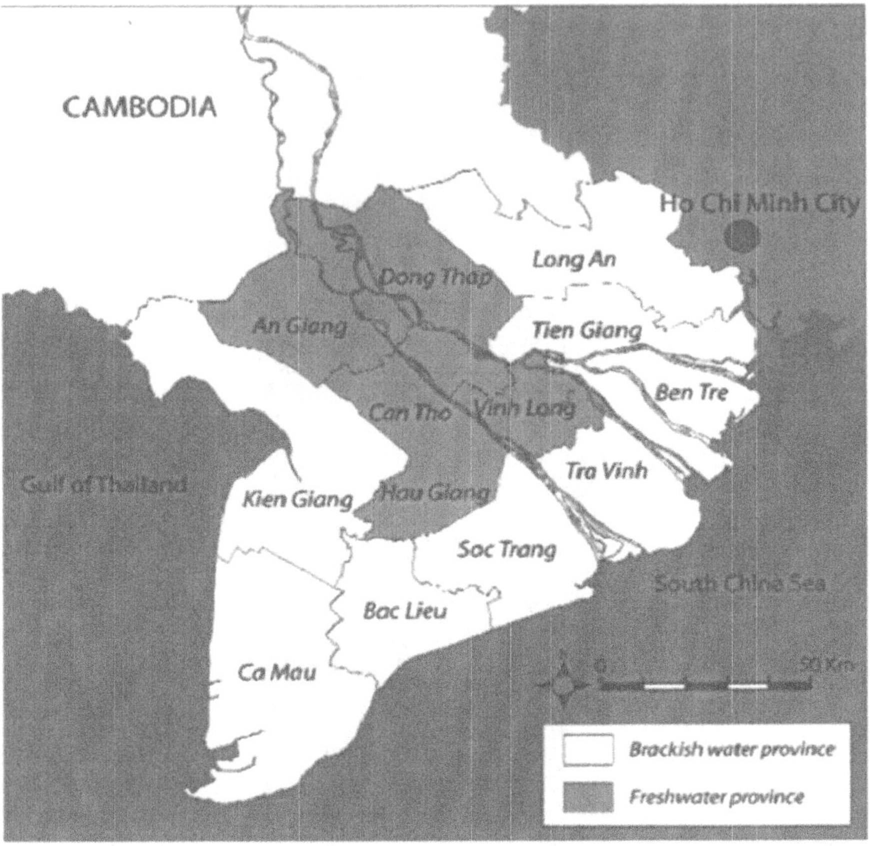

Fig. 7.1 Main catfish farming areas (shaded) in the Mekong Delta

7.4 The Role of Striped Catfish in Aquaculture Sector

Recently, catfish has begun to play a very important and significant role in the aquaculture sector of Vietnam. The production of catfish in 2007 was 1,200,000 tons, accounting for more than 50% of the total aquaculture production of Vietnam (Fig. 7.3). This production includes 95–97% striped catfish, and 2–3% Mekong river catfish. Striped catfish, therefore, has become the beautiful "Princess in Vietnamese Aquaculture" in recent years. The growth of striped catfish farming has been dramatic, increasing many fold to reach 1,200,000 tons in the 10-year period from 1997 to 2007. Farming areas increased about 8-fold from 1,250 ha to over 9,000 ha, while the production increased 45-fold from around 22,500 to 1,200,000 tons (Dung 2008).

As such, striped catfish will continue to be the key species in Vietnamese aquaculture, and will have a strong impact on the success of the whole aquaculture sector of the country.

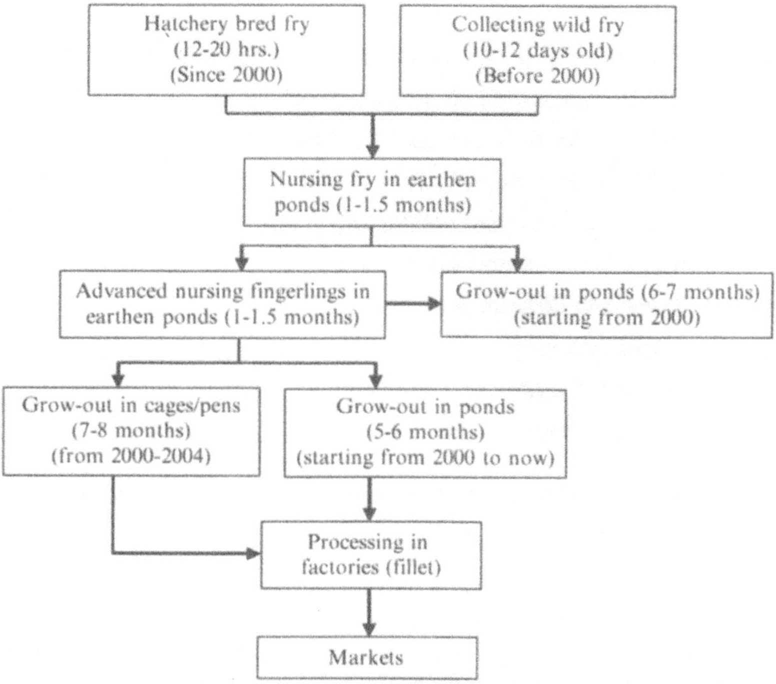

Fig. 7.2 Production chain of striped catfish farming in Vietnam

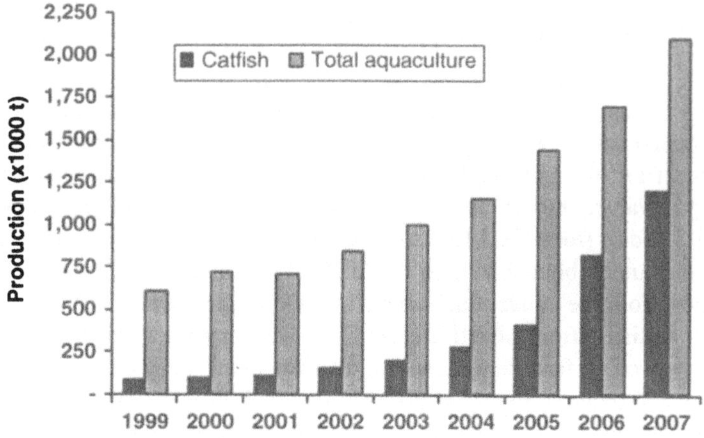

Fig. 7.3 Production of striped catfish in comparison to total aquaculture production in Vietnam (compiled from Phuong 2007; Dung 2008)

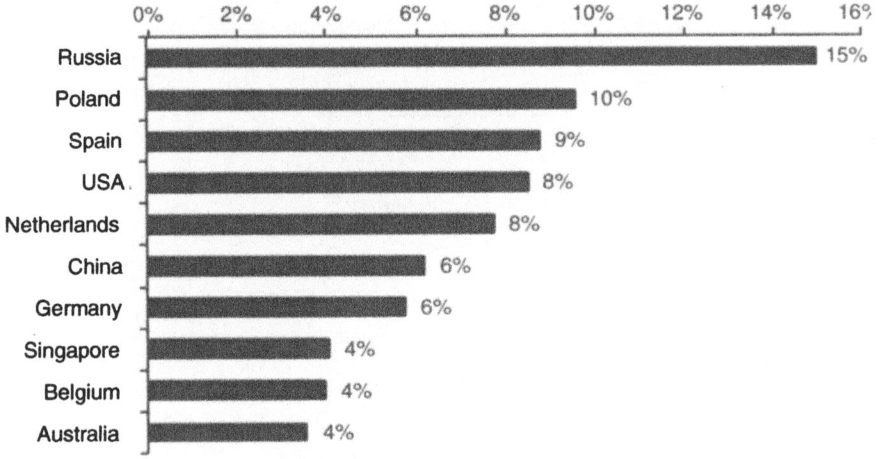

Fig. 7.4 Ten leading striped catfish importing countries from Vietnam in 2006 (from Dung 2008)

Besides, the processing infrastructure for striped catfish was also developed rather fast to meet the processing demands. Within a 10 year period (1997–2007), 80 processing plants with a combined capacity of up to 3,500 tons raw fish daily were established, mostly in the Mekong delta. This resulted in an increase of over 55-fold, from 7,000 to 386,870 tons the volume of exported fillets. Striped catfish products have also been exported to over 80 countries and territories (Dung 2008) (Fig. 7.4). Although, striped catfish has been exported mostly in fillet form, there are more than 40 value added products have also been produced for domestic and international markets.

7.5 Socio economical Impacts from Striped Catfish Farming

The socio-economic impacts of the striped catfish farming may need a detail study for full understanding. However, the direct impact on the economic aspect can be represented from the household income of striped catfish farmers and increased yearly export value. During the 10-year period from 1997 to 2007, the export values increased 50-fold, from US $19.7 to 979,036 (Fig. 7.5). The amount earned from export contributed about 0.84% to the GDP of Vietnam or 21.7% of 3.7% of the contribution from the aquaculture sector in 2006 to the GDP.

Striped catfish farming itself creates livelihood opportunities not only to those directly involved in farm activities, but also for many other people who work as labor in related sectors. Striped catfish farming activities initiated the development of fillet processing, feeds, drugs, and chemicals industries. These related industries have provided livelihood opportunities for a large number of people, a significant proportion of those are rural women.

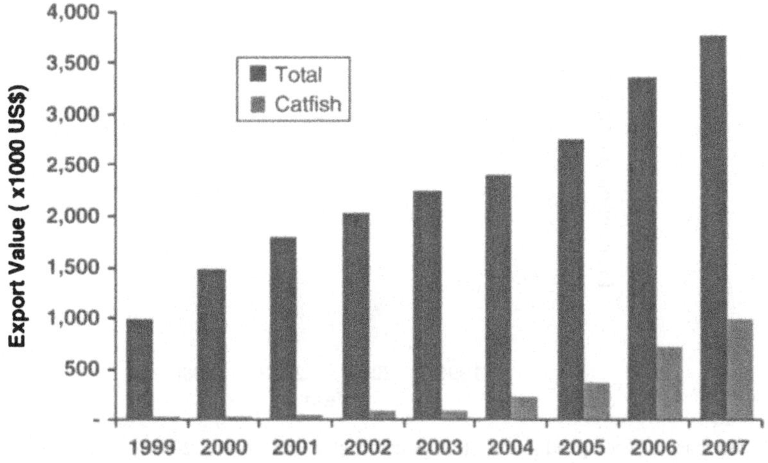

Fig. 7.5 Export value of catfish in comparison with total export value from aquaculture and fisheries sector of Vietnam (Dung 2008)

7.6 Key Factors of for the Success of Striped Catfish Farming in Vietnam

7.6.1 Success of Seed Production and Hatchery Development

The availability of good quality seed stock in adequate quantities is considered as one of a few key drivers to the explosive growth of striped catfish culture in Vietnam. Before 2000, the seed stock was wild caught and of limited quantity and seasonal availability. Tung et al. (2001) (cited by Van Zalinge et al. 2002) estimated that the annual amount of wild caught fry varied from 50 to 200 millions during the period from 1977 to 1997. The culture of striped catfish in that period was mainly in small ponds with low stocking density. However, since seed was produced year-round in hatcheries in adequate amounts, the commercial culture was developed with multiple crops per year and enabled the adoption of extremely high stocking densities.

Research on artificial propagation of Pangasiid catfish was first commenced in 1978 (Xuan 1994) with *P. hypophthalmus* species. However, results were not sufficiently reliable for mass seed production and the research activities were discontinued. Then research on induced spawning of striped catfish was restarted in 1995 under an EU funded project, which was led by Can Tho University. Partners of this project included French Agricultural Research Centre for International Development (CIRAD), Research Institute for Development (IRD) (France), Can Tho University (CTU) and An Giang Fisheries Import–Export joint stock (AGIFISH) Company (Vietnam). Techniques for induced spawning of striped catfish were primarily

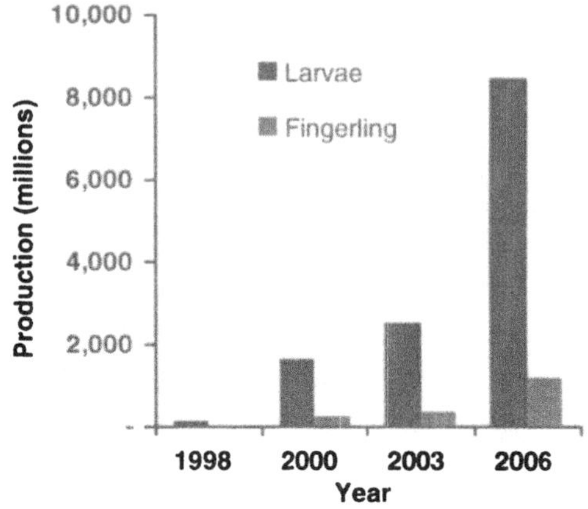

Fig. 7.6 Increase of seed production of striped catfish from 1998 to 2006

established in 1996 and fully achieved and transferred to hatchery operators since 2000 (Cacot 1999; Cacot et al. 2002). Since then, the seed production of striped catfish has been increased rapidly (Fig. 7.6).

In the early stages, striped catfish seed production was only conducted in four state-run hatcheries, and the production did not meet the demand for stocking. Since 2002, a number of private hatcheries have been established with technical support from CTU and Research Institute for Aquaculture No. 2 contributed significantly to the expansion of striped catfish farms. Striped catfish hatcheries are now capable of meeting the seedstock demand for commercial grow-out in the region (Fig. 7.6). However, due to the increasing demand for high quality striped catfish fingerlings for grow-out, a number of large-scale grow-out farms have built hatcheries to produce seedstock for their own use. This trend has continued to this day.

7.7 Success of Pond Culture Development

7.7.1 Overview of Culture Systems

Three main culture practices of striped catfish occur in the Mekong Delta; pond, cage, and pen. The development of these types has changed with time according to its economic efficiency. Pangasiid catfish culture in cages was introduced and developed very early in the Delta. At the beginning of the 1960s, a number of Vietnamese who had lived in Cambodia and returned to Vietnam commenced catfish cage culture practice in the border area of Vietnam and Cambodia (Interim Committee for Coordinating of Investigations of the Lower Mekong 1992; Phuong 1998). Mekong river catfish was the main culture species in cages in that period.

At the beginning of the year 2000, Mekong river catfish was replaced by striped catfish to be cultured in cages when seed stock supplies became freely available from developments in hatchery production of this species. Whereas, seedstock of Mekong river catfish species was still dependent on wild stocks, and thus, could not meet the demand for grow-out culture.

The cages used in striped catfish culture were of three sizes. Small-sized was less than 288 m^3, medium-sized ranged from 288 to 720 m^3 and the largest was above 720 m^3. The medium-sized comprised 56.7%, and large size comprised 36.7% of total cages (Nhi 2005). Pen culture, which is a fixed enclosure built on the river bank with metal or bamboo net, also developed since 2000. This model was comparable to that used traditionally for the culture of giant freshwater prawn (*Macrobrachium rosenbergii*). The average pen size is around 3,200 m^2. However, cage and pen cultures of striped catfish that commenced in 2000 lasted only for a short time. By 2004, these culture practices were significantly reduced and became unimportant in striped catfish culture. The decline of these culture practices was primarily due to slower fish growth, higher fish mortality, and frequent disease outbreaks that lead to reduced economic efficiency compared to pond practices (Phuong et al. 2004).

Catfish culture in ponds, however, has been developing very rapidly and reached the highest production share, during the last few years. Dung (2008) reported that by 2007, striped catfish was cultured in an area of 9,000 ha and the production was almost 1,200,000 tons, which comprised more than 95% total of striped catfish production (Fig. 7.7). The important driving forces in the development of pond culture practice include the initiative in seed supply all year-round, low infrastructural investment, short culture period, and high economic efficiency. At present, catfish culture in ponds is still developing fast. The farm size is considered as small scale. The study of Hien (2008) indicated that 57.3% of farms have one pond,

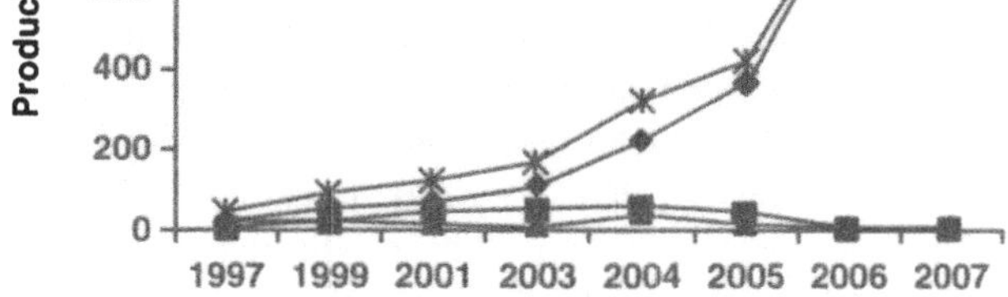

Fig 7.7 Share of production of different striped catfish culture systems (Compiled from Mien 2004; Phuong 2007; Dung 2008)

36.2% have 2–4 ponds, and 6.5% hasve more than 4 ponds. Liem et al. (2008) reported that some large farms may have up to 26 ponds.

7.7.2 Key Culture Techniques of Pond Culture

The first striped catfish ponds were built on alluvial deposits along the main branches of the Mekong River of good soil quality and plentiful supply of water. Farms are normally designed with fixed supportive systems, such as water supply and drainage systems, feeding stations, and water pumps. A good farm design must ensure possibility of a high water exchange for ponds, especially at the end of the culture cycle. Pond size is quite similar around 0.4–0.42 ha. Taking advantage of a plentiful water supply, ponds are designed with high water level of 3.6–4 m for high stocking density. Phuong et al. (2004) reported that the average stocking densities of striped catfish in pond were was 20.5 fish m^{-2}, and it has been increased remarkably up to 52.8 fish m^{-2}, especially in some cases up to 75 fish m^{-2} (Liem 2009). The productivity is, therefore, relatively high and kept improving. The average productivity of pond culture was 20.5 kg $m^{-2}crop^{-1}$ in 2005 (Nhi 2005) and then 36.9 kg $m^{-2}crop^{-1}$, the highest was greater than 500 kg m^{-2} in 2008 (Hien 2008). High productivity of striped catfish culture in ponds has been due to the appropriate exploitation of the biological features of this air-breathing species in that they are good survivors in condition of high stocking density. However, the vital factor for high productivity of striped catfish culture in pond is the improvement of culture techniques, such as pellet feeding, deep ponds, and high water exchange regimes.

7.7.3 Improvement of Feeds

Improvement of feeds is one of the key factors for the success of striped catfish farming. Feeds used in striped catfish farming have changed rapidly in the past years, especially after 2000 from low quality to nutritionally balanced feeds. During the early developing years of catfish culture, farm-made feeds (sinking form) were mainly used. Phuong (1998) reported that in 1995, almost 100% catfish farmers used low quality farm-made feeds prepared using by products from agriculture and fisheries. However, striped catfish has gradually been fed pelleted feeds (floating form). Liem et al. (2008) reported that the use of feed types for striped catfish varied by farm sites, 50–88% farms used pelleted feeds only and 13–40% farms a combination of pelleted and farm-made feeds. This indicates that striped catfish farming in the Mekong delta has gradually been moved from using farm-made to pelleted feeds. However, because of the lower cost, the use of farm-made feeds may continue. A study of Phuong (2007) indicated that striped catfish pond culture using farm-made feeds gave higher net profit compared to commercial pellet. The quality of recent farm-made feeds has improved significantly in which dried low value fishes and soybean meal are used to improve its protein content. Nhi (2005) reported that the protein content in farm-made feed for pond culture varied from 22.2 to 29.5% (dry weight basis) compared to 17.9–22.6% in commercial pellet feeds (Phuong 1988). It is hard to determine the exact amount of pelleted feeds used for striped catfish culture in the Mekong Delta, but an approximate estimation is that 1.2–1.5 million t were used in the year 2007.

7.7.4 Advantage of Water Sources

Striped catfish culture is based on the advantage of the availability of a plentiful water supply from the Mekong River. The environmental management of striped culture models is largely dependednt on river water. Mekong is a large river with high water flow, especially during flood season that a large water volume converge from upper reaches and sweeps away waste matters from cages and pens, and this is a noticeable advantage contributing to the development of these culture models. Pond culture environment has total ammonia nitrogen (TAN), nitrite (NO^{2-}), and phosphorus (PO_4^{3-}), BOD and H_2S much higher than the recommended levels (Giang et al. 2008) due to high stocking density. However, high rate of water exchange is the main method applied to improve pond water quality. In theory, treatment is required for water supply and waste water, but this is absolutely impossible in practice. Striped catfish ponds contain a large volume of water and land used for catfish ponds is very expensive, and thus, it is impractical and inapplicable to have settlement and treatment ponds in catfish farms.

7.7.5 High Economic Returns

High economic return is an attractive reason for the involvement of investors in catfish farming. Most recent striped catfish farmers are investors from other sectors. Striped catfish farming requires high level of investment for land, infrastructural building, and operating costs (mainly feed and seed) (Table 7.1). The production cost and net profit per unit of fish harvest are relatively lower than for other species. However, total net profit per ha is much higher than other farmed species because of higher productivity. The production cost has been increasing over time. Phuong et al. (2007) reported that the production cost (US $/kg) in 2006 was 0.59, which is lower than that of 0.70 in 2008 (Hien 2008). Hien (2008) also stated that 11.6% of farms encountered a financial loss in 2008. The losses in catfish farming recently have been mainly due to the reduction of prices.

7.7.6 Traditional Practices and Farmer Innovations

As mentioned earlier, striped catfish culture has a long historical practice in the Mekong Delta. The farmers in Dong Thap and An Giang Provinces, which are close to the Cambodian border, have a long tradition of catching wild fry and nursing

Table 7.1 Summary of assessed financial and economic indicators of pond culture (per ha) (Hien 2008 in US $)

Item	Total	Percent share
Total costs	230,188	
Total fixed cost	1,263	
Total variable cost	228,952	
Feed	166,219	72.6
Seed	16,027	7.00
Chemicals and drugs	9,845	4.30
Interest	5,724	2.50
Power (gasoline, electricity,...)	1,603	0.70
Labors	3,663	1.6
Other costs	25,872	11.3
Gross income	274,394	
Net income	44,206	
Production cost per kg	0.696	
Net income per kg	0.134	

fingerlings for distribution to farmers in other areas of the delta to stock in small ponds. Striped catfish culture in ponds has become a universal practice of many farmers in the region and the farmers are also knowledgeable about the fish and its culture techniques. With this advantage, the Vietnamese farmers have adapted seed production and grow-out technologies easily and rapidly. Moreover, the striped catfish farmers are also very innovative in improving farming technologies. These characters of farmers have contributed greatly to the fast growth of the industry.

7.7.7 Government Support

The Vietnamese government has recognized the role and potential of striped catfish species in the context of aquaculture development. Therefore, different supportive policies from government have been established overtime. Technical trainings and consultation programs for catfish farmers have been intensively implemented via universities, research institutions, and local organizations, and these have played a vital role during the early developmental stages of the industry. The government has established other forms of support such as bank loan mechanisms for both producers and processors to expand production scale and to establish and/ or improve processing infrastructures. International trade promotion programs have also been supported by the government in conjunction with entrepreneurial activities. Finally, the government has also assisted with various scaled research projects related to striped catfish farming, such as seed quality improvement, disease control, and environmental monitoring.

7.7.8 *Availability of Markets*

The availability of markets, especially export markets, has also been one of the key factors contributing to the fast growth of striped catfish industry. Markets for striped catfish fillet have been expanded over time. Striped catfish has become an acceptable substitute for "white flesh" fish, such as cod, for the western palate. Prior to 2003, export markets for striped catfish were very limited; the US was the most important one. Recently, striped catfish fillet has been exported to over 80 countries world wide. Of which, the market share was 48% in European countries such as the Netherlands and Germany, 9.2% in Russia (a new market) and only 6.9% in the US (a traditional market) (Dung 2008). The rapid expansion of new markets, together with product quality improvement (traceability and food safety), played a key role for the increase of production in the recent past.

7.8 Major Challenges in Up-scaling of Striped Catfish Industry in Vietnam

7.8.1 *Diseases, Drugs, and Chemical Uses*

Diseases in striped catfish farming have been one of the major obstacles to its sustainable development. The intensification process of the culture practices has resulted in disease outbreaks, particularly during the last few years. Several pathogens have been reported to be associated with diseases in striped catfish culture, including fungi, ento- and ecto-parasites as well as bacteria. Among the infectious diseases, bacterial agents have been responsible for the major epizootics affecting striped catfish farming. Rate of loss due to bacterial diseases in striped catfish farming (such as white spots on the internal organs caused by *Edwardsiella ictaluri*; hemorrhagic symptoms caused by pathogenic *Aeromonas hydrophilla,* and collumaris disease caused by *Flexibacter collumnaris*) has been estimated as high as 50% compared to others (Thinh et al. 2004; Phuong et al. 2007). The outbreak of diseases has always been primarily caused by stress induced factors, and is of a seasonal nature. Studies have revealed that stress factors that include localized environmental degradation, wastes from agricultural activities, improper caring, high stocking density, and low quality of stocked seed make the stock susceptible to infectious pathogens (Phuong et al. 2007).

Disease outbreaks and the consequent improper use of drugs and chemicals have been considered as one of the major challenges in sustainable growth of striped catfish farming sector. The use of chemicals and drugs is very common in catfish farming and relates to food safety. Liem (2008) found that there are 39 varieties of chemicals and drugs used for water quality regulation, disease treatment and prevention. Antibiotics use to treat bacterial diseases is of high concern due to possible residues in fish flesh. The use of antibiotics has reduced over time to comply with the food safety standards required by importing markets. In fact, the list of banned or limited use of antibiotics and chemicals used in aquaculture in Vietnam is updated regularly. Currently, only a few antibiotics are permitted to be used in catfish farming.

7.8.2 Environmental Sustainability

The sustainability of striped catfish farming is very well related to external aquatic environment. High amount of effluents from striped catfish ponds that cannot be treated have been considered as a source of pollution causing degradation of water quality in the river. Although, striped catfish farms are mostly located along two branches of the Mekong River, which helps to carry wastes to minimize pollution, but localized pollution of water has been recorded in intensive farms. In certain areas, higher mortality of stocked seed and lower survival rate at harvest in comparison to previous years have been observed, and are thought to be associated to with poorer water quality. Therefore, control of effluent water quality from striped catfish pond culture systems to minimize environmental pollution is a crucial issue to enable sustainable development of striped catfish culture in the Delta. Efforts to minimize environmental pollution, such as farm registration, approval of new farm establishment, demonstration and encouragement to apply good aquaculture practices (GAPs), have been attempted through management schemes from local authorities. Furthermore, recommendations on reducing stocking density to reduce pollution and associated risks have also been introduced to catfish farmers.

7.8.3 Seed Quality

The increase of seed demand has created concerns on seed quality. Striped catfish seed quality is highly influenced by the farmers' knowledge on broodstock quality management in hatcheries. Several studies have shown that brooders are induced to spawn many times each year, the rate that brooders are added/changed to the broodstock population is relatively low, brooders often originate from the same source/family (appearing in 90% hatcheries); and cross breeding between males and females are undiversified. Hien (2008) reported that 21.1% of grow-out farmers found that the quality of seedstocks seems to have declined over the years.

7.9 Ensuring Sustainable Striped Catfish Farming Systems

The striped catfish farming has shown some factors that could impact on its sustainability and potential pollution from waste water and effluents, disease outbreaks, reduction of seed quality being foremost. It is urgently needed to develop and implement sustainable production systems with best management practices (BMP) and GAP. Fortunately, there have been various pilot projects to implement sustainable production models, such as the demonstration of SQF-1000 (Safe & Quality Food) that started in 2005, GAP in 2007, and BMP project funded by ACIAR that is being conducted jointly by the Network of Aquaculture Centres in Asia-Pacific (NACA), Department of Primary Industries, Victoria, Australia, Research Institute for Aquaculture No. 2, and Can Tho University.

7.9.1 Markets

The export markets for Vietnamese catfish products have been diversified into more than 80 countries and territories. However, the striped catfish industry in Vietnam has faced new challenges that include increased requirements on food safety and traceability, and taxation from importing markets. These could limit the export of the striped catfish products. The future of striped catfish products requires a proper policy in widening export, expanding local and regional markets, and identifying market demand in order to balance the production. Moreover, the development of brand image for striped catfish products from Vietnam is very important to retain the existing and establish new markets.

7.9.2 Organization of the Industry

The existing status of striped catfish production chain is not well organized. The linkage among stakeholders or main actors, especially farmers, processors and suppliers (feeds, drugs, chemicals, and others), bankers, etc., has been rather weak or without an apparent linkage. Farmers are much more vulnerable in the production chain. Therefore, a vertical linkage among main actors must be established urgently in order to share information, responsibility, and benefits.

7.10 Conclusions

The striped catfish farming plays a vital role in the aquaculture sector in Vietnam, and is in a fast growth phase in terms of culture area and production. Equally, it is important that this sector provides an acceptable "white fish" to the Western world, in place of the traditional white fish products such as cod, at a very reasonable price. This species will continue to be a key aquatic export commodity of Vietnam and contribute increasingly to Vietnamese aquaculture production and to the economy, the great bulk of which will trickle down to the poor rural communities of the Mekong Delta. The success of this industry can be attributed to advantages of natural conditions and accumulated efforts from research, trading, infrastructural development, and government support. Though many issues need to be addressed, it is believed that this sector can sustain and the striped catfish will remain as an unprecedented development of aquaculture world-wide.

References

Cacot, P. 1999. Description of the sexual cycle related to the environment and set up of the artificial propagation in *Pangasius bocourti* (Sauvage 1880) and *Pangasius hypophthalmus* (Sauvage 1878) reared in floating cages and in ponds in the Mekong Delta. In M. Legendre and A. Pariselle (eds.). *The biological diversity and aquaculture of Clariid and Pangasiid catfishes in South East Asia*, Proceedings of the mid-term workshop of the "Catfish Asia Project" Can Tho, Viet Nam, 11–15 May 1998, pp. 71–89. Toulouse, France: Paragraphic.

Cacot, P., M. Legendre, Q.D. Tran, T.T. Le, P.T. Liem, C. Marojouls, and J. Lazard. 2002. Induced ovulation of *Pangasius bocourti* (Sauvage 1880) with a progressive HCG treatment. *Aquaculture* 213: 199–206.

Dung, N.H. 2008. Achieving a sustainable future for Vietnamese seafood industry. Keynote Speech at the IIFET 2008 Conference, Nha Trang, Vietnam, July 22–25, 2008.

Giang, H.T., V.N. Ut, and N.T. Phuong. 2008. Study on water quality of intensive catfish culture (*Pangasianodon hypophthalmus*) ponds in An Giang province. *Scientific Journal of Can Tho University* (Special Issue on Aquaculture and Fisheries) 1: 1–9 (in Vietnamese).

Hien, L.L. 2008. Analysis of the supply and demand of Striped catfish (*Pangasianodon hypophthalmus*) in the Mekong river delta. Master Thesis in Aquaculture. College of Aquacultrue and Fisheries, Can Tho University, Vietnam, 130 (in Vietnamese).

Interim Committee for Cordinating of Investigations of the Lower Mekong. 1992. Fisheries in the Lower Mekong Basin, Annex 4 and 5, Bangkok.

Liem, P.T., N.V. Phong, and N.T. Phuong. 2009. The use of drugs and chemicals in catfish farming in the Mekong River Delta, Vietnam: A review. *Scientific Journal of Can Tho University*, Viet Nam (in Vietnamese).

Mien, T.T. 2004. *Issues on planning and management for catfish production to the year 2010.* Paper Presented at the National Conference on Quality and Trade Name for Catfishes of Viet Nam. An Giang Province, December 14–15, 2004, 14 (in Vietnamese).

Nhi, T.V. 2005. The use of local feedstuffs for Striped catfish (*Pangasianodon hypophthalmus*) feed formulation in An Giang province. Master thesis of Aquaculture. College of Aquaculture and Fisheries, Can Tho University, Vietnam, 59 (in Vietnamese).

Phuong, N.T. 1998. Pangasius catfish cage culture in the Mekong Delta: Current status and study for feeding improvement. PhD Thesis. Institut National Polytechnique de Toulouse, France 134 pp.

Phuong, N.T., and T.H. Minh. 2005. National aquaculture sector overview – Viet Nam. In FAO Fisheries and Aquaculture Department [online]. Rome. Updated 10 October 2005. [Cited 23 July 2008]. http://www.fao.org/fishery/countrysector/naso_vietnam/en.

Phuong, N.T., P.M. Duc, V.N. Son, T.V. Bui, and A.T.A. Nguyet. 2004. A review of the application of bio-technologies for quality improvement and production cost reduction of giant freshwater prawn (*Macrobracium rosenbergii*), catfishes (*Pangasius bocourti* and *Pangasius hypophthalmus*) and Tilapia (Oreochromis niloticus) farming in An Giang province. Report submitted to the Department of Science and Technology of An Giang province, 24 (in Vietnamese).

Phuong, N.T., L.X. Sinh, N.Q. Thinh, H.H. Chau, C.T. Anh, and N.M. Hau. 2007. Economics of aquaculture feeding practices: Viet Nam. *FAO Fisheries Technical Paper* 505:183–205.

Thinh, N.Q., T.D. Dung, and H.W. Ferguson. 2004. Histopathology of white spot in internal organs of Pangasius hypophthalmus. *Scientific Journal of Can Tho University* (Special Issue on Aquaculture and Fisheries) 120–125 (in Vietnamese).

Tien V., 1993. Fisheries development in the Mekong delta, Vietnam. In *Fisheries magazine*, Ministry of Fisheries, 22–30 (in Vietnamese).

Trong, T.Q., N.V. Hao, and D. Griffiths. 2002. Status of Pangasiid aquacultrue in Viet Nam. MRC Techincal Paper No. 2. Mekong River Commission, Phnom Penh, 16 pp.

Tuan, N.A., N.T. Phuong, P.T. Liem, and N.V. Thuong. 2003. Results of the study on Pangasius catfishes and their future development. *Journal of Mekong Fisheries*. pp. 129–134 (in Vietnamese). Agricultural publishing house. Ho Chi Minh city. Viet Nam.

Van Zalinge, N., L. Sopha, N.P. Bun, H. Kong, and J.V. Jorgensen. 2002. Status of the Mekong *Pangasianodon hypophthalmus* resources, with special references to the stock shared between Cambodia and Viet Nam, 29. Phnom Penh: Mekong River Commission.

Xuan, T.T. 1994. *Some biological characteristics and artificial reproduction of river catfish (P. micronemus Bleeker) in the South Vietnam*. Paper Presented at the International Workshop on the Biological Bases for Aquaculture of Siluriformes, May 24–27/1994, Montpellier, France.

Chapter 8
The Genetic Improvement of Farmed Tilapias Project: Impact and Lessons Learned

Belen O. Acosta and Modadugu V. Gupta

Abstract In response to challenges that the developing world confront on food security and malnutrition, the last two decades have witnessed increased efforts in genetic improvement to enhance production traits of commercially important aquatic species. From the 1980s to the present, several institutions in developing countries have been engaged in such R&D activity and it is recognized that the collaborative program on Genetic Improvement of Farmed Tilapias (GIFT) has spurred the development of several tilapia and carp breeding programs that now exist in numerous developing countries. The GIFT is a collaborative R&D program conducted by the WorldFish Center (formerly, International Center for Living Aquatic Resources Management, ICLARM) and its partners from the Philippines and Norway aimed to develop methodologies for the genetic improvement of tropical finfish of aquaculture importance. The GIFT project has demonstrated that selective breeding is a feasible, cost effective, and sustainable approach to the genetic improvement of tropical finfish, and also confirmed the importance of a multidisciplinary approach that enabled the assessment of economic viability, social acceptability, and environmental compatibility, thus, creating confidence among planners and administrators, all of which facilitated the transfer of research findings to farming systems in a host of countries. The program and its successors, such as the International Network on Genetics in Aquaculture (INGA), demonstrated that networking and partnership building among national institutions in developing countries, advanced scientific institutions, and regional and international organizations can play a major role in accelerating research and the success of R&D.

B.O. Acosta (✉) and M.V. Gupta
Aquaculture Department, Southeast Asian Fisheries Development Center Manila Office, 102 Philippine Social Science Center Building, Commonwealth Avenue, Quezon City Diliman, 1101, Philippines
e-mail: bacosta@seafdec.org.ph

S.S. De Silva and F.B. Davy (eds.), *Success Stories in Asian Aquaculture*,

8.1 Introduction

8.1.1 Background

Fish and fisheries have been playing an important role in addressing nutritional and livelihood security, especially of the poor in the developing countries. Globally, over 2 billion people get at least 20% of their animal protein intake from fish. In industrialized countries, fish provide 13% of animal protein intake, whilst in Asia, it averages 30%. In some Asian countries, it could be much higher than the average: 51, 58, and 75% in Bangladesh, Indonesia, and Cambodia, respectively (Delgado et al. 2003). In addition to contributing to nutritional security, aquaculture has been providing livelihood to millions of people across the world.

As against a production of 107 million tons of food fish, it has been estimated that by the year 2030, an additional 30–40 million tons of food fish would be needed to meet the demand from the growing world population, changing dietary habits, and increasing income levels (Delgado et al. 2003; FAO 2007a).

Global fish production from capture fisheries in the last decade has stagnated, and probably no major increases can be expected. FAO (2007b) estimated that 52% of the marine fish stocks are fully exploited. Studies undertaken in the 1990s in nine countries in Asia (Bangladesh, China, India, Indonesia, Malaysia, Philippines, Thailand, and Vietnam) indicated that inshore demersal fish populations that constitute a major portion of coastal fisheries and constitute over 80% of the total fish production have declined by 4–44% when compared with what they were in early the 1970s (Silvestre et al. 2003).

In the abovementioned scenario, the increasing demand has to be met from aquaculture. Aquaculture has been the fastest growing food production sector in the last three decades, with an annual growth of about 8.8% during 1950–2004 (FAO 2006a). While there is still an enormous potential for increasing production through aquaculture, these production increases have to be sustainable without impinging on the environmental integrity. While productivity could be increased to an extent through improved farm management practices, a significant component of the increases in production required to meet the global demand have to come from the use of genetically improved strains/breeds of fish, as has been for crops and livestock. The Green Revolution of the 1960s and 1970s, with its package of genetically improved seeds and farm technology has been responsible for the enormous increases in crop yields in many developing countries. At least 30% of the increase in the production of protein has resulted from genetic research and comprehensive breeding programs (FAO 2001). The success of genetic improvement programs for salmon and trout in Norway is similar to that achieved in livestock and poultry.

In view of the increasing expectation from fish farming to help address food security and promote rural development, significant progress has been made in the last two decades in tropical aquaculture. For instance, in the case of tilapias, there is of late an emergence of the better growing strains used for aquaculture which is the result of significant R&D efforts on genetic improvement. From the 1980s to the present, several institutions in the developing countries have been engaged in this R&D activity and it is recognized that the collaborative program on Genetic Improvement of Farmed Tilapias (GIFT) has spurred the establishment of tropical fish breeding programs and the development of improved strains (mostly tilapias and carps) that are now available in many parts of Asia. The GIFT is a collaborative R&D program which was conducted to develop methodologies for the genetic improvement of tropical fish species using the Nile tilapia (*Oreochromis niloticus*) as the test species.

8.1.2 Objectives of the Paper

This paper is primarily intended to document the lessons learned from the GIFT program and the elements that have contributed to its success. Specifically, it aims to:

- Highlight the experiences in the development of GIFT Program which have led to positive societal change and negligible ecosystem impacts
- Provide a better understanding of the approaches used in the GIFT Program that could contribute to sustainable aquaculture growth combined with positive societal change
- Provide a direction for future policy change and collective improvement to contribute to sustainability of the aquaculture sector through genetic improvement
- Mitigate negative public perceptions on the improved strain of fish, as exemplified by the tilapia and the associated genetic improvement technology used

8.1.3 Structure of the Document

This paper highlights the experiences, impacts, and lessons learned in the GIFT program in six steps:

- Explains the importance of genetic enhancement in fish and the use of Nile tilapia (*O. niloticus*) as a test species for developing the technology for genetic improvement of tropical finfish
- Traces the evolution of the program for the genetic enhancement of Nile tilapia, and consequently, the development and dissemination of the improved strain with a focus on how the implementing institutions ensured the appropriateness and suitability of the approaches used
- Discusses the efforts made to create an environment of sustainability in the GIFT program
- Describes the development of a global initiative on genetic improvement, and how this initiative has created an impact on the aquaculture industry and the national economy in general
- Enumerates the lessons learned in the GIFT program which could help provide a direction for future policy change and collective improvements in the aquaculture sector through genetic improvement

8.2 Why is the Need for Genetic Improvement of Fish?

The green revolution of the 1960s and 1970s, with its package of genetically improved seed, farm technology, and better farm inputs, has been responsible for the enormous increases in crop yields in many developing countries (FAO 2001). While significant production gains have been made in the case of crops, livestock, and temperate fish such as salmon, no efforts were made until the 1980s for the development of improved strains or methods for genetic improvement of tropical finfish, which contribute to over 50% of the global aquaculture production. It has been estimated that less than 1% of the global aquaculture production comes from genetically improved stocks, indicating enormous potential for production gains through the use of improved strains in aquaculture. Further, it has been shown that in many cases, the cultured stocks are inferior to the wild populations due to inbreeding (Pullin and Capili 1988).

In view of the above mentioned facts, the WorldFish Center, in collaboration with the national partners in Philippines and Norway, initiated a research in 1988 to develop methods for the genetic improvement of tropical finfish, using tilapia as a test species.

8.3 Why Tilapia?

Tilapias are farmed in at least 85 countries, because of many desirable qualities such as the ability to survive and grow in shallow and turbid waters, suitability for culture in low-input extensive systems to high-input intensive systems, and comparatively higher resistance to diseases and parasites when compared with other aquaculture species (Pullin 1983; Eknath 1995; Gupta and Acosta 2001a), with most production from the developing countries of Asia and Latin America. In aquaculture, tilapias are regarded as opportunistic omnivores and herbivores feeding on phytoplankton and detritus, just like the world's principle farmed livestock (Beveridge and Baird 2000). It is considered as a poor man's fish and/or an "aquatic chicken." With increasing popularity among consumers, tilapias have become the world's second most popular farmed fish, after carps. Global production of farmed tilapia reached 2.3 million tons in 2006, valued at about $2.4 billion (Fitzsimmons 2008). Because of the abovementioned qualities and short generation time of about six months and its suitability for the application of genetics in aquaculture, from conservation of genetic resources to breeding programs, tilapia was considered as an ideal species for developing methods for genetic improvement of tropical fish.

Studies conducted by the WorldFish Center and its partners during the 1980s found that Asian tilapia stocks were of poor genetic quality and this was largely attributed to poor broodstock management, resulting in inbreeding and widespread introgression of genes from undesirable feral *O. mossambicus* (Macaranas et al. 1984; Macaranas et al. 1986; Taniguchi et al. 1985; Eknath et al. 1991). A review of the history of introductions and subsequent transfers of *O.niloticus* also revealed that the stocks in Asia were descendants of a few introductions consisting of very few fish and were probably suffering from genetic bottlenecks (Pullin and Capili 1988).

8.3.1 Emergence of Partnership for Genetic Enhancement of Tilapia

The findings described earlier which were based on extensive research and backed by surveys and consultations with various stakeholder groups (see Fig. 8.1) led to a consensus that tropical finfish used for aquaculture was in dire need of genetic improvement.

Consequently, WorldFish Center and its partners [Norwegian Institute of Aquaculture Research – AKVAFORSK and Philippine national fisheries institutions (Bureau of Fisheries and Aquatic Resources, BFAR; Freshwater Aquaculture

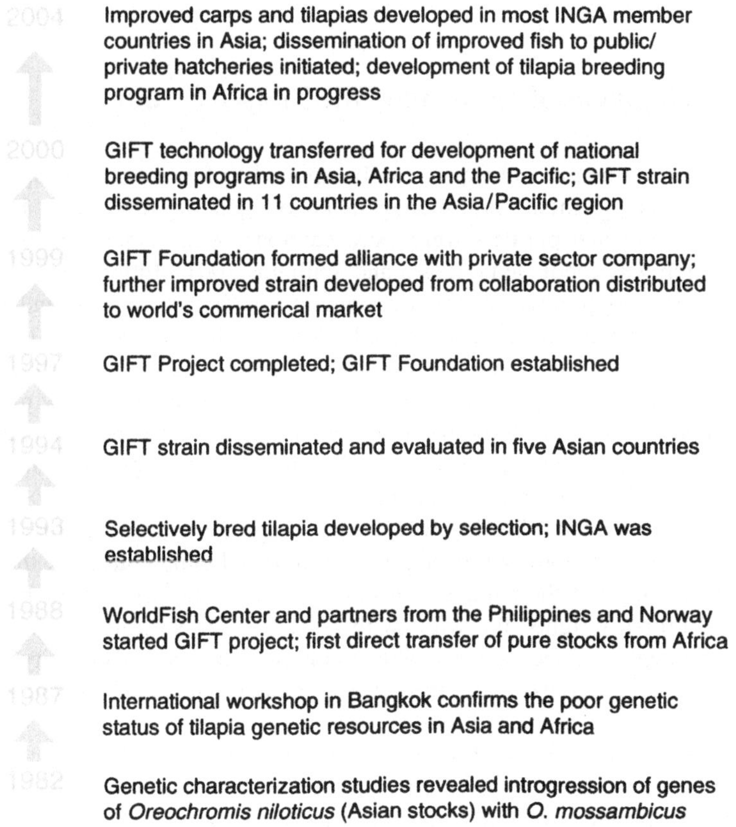

Fig. 8.1 The Evolution of the Genetic Improvement of Farmed Tilapias (GIFT) program and globalization of research outputs and methods (source: Gupta and Acosta 2004)

Center of the Central Luzon State University, FAC-CLSU; and the Marine Science Institute of the University of the Philippines, UPMSI)] pursued a collaborative program for the genetic improvement of tropical finfish using the tilapias as the test species in 1988. This program which came to be known as the GIFT project was conducted for over almost 10 years and addressed the following objectives: to (1) develop methods for genetic enhancement of tropical finfish; (2) develop improved breeds of Nile tilapia (*O. niloticus*); (3) build capacity of national institutions in aquaculture genetics research; (4) disseminate improved strain (GIFT strain); (5) carry out genetic, socioeconomic, and environmental evaluation of GIFT; and (6) facilitate development of national tilapia breeding programs (Eknath 1994).

Apart from the main partners, the project also established linkages with national institutions in Egypt, Ghana, Kenya, and Senegal, the University of Hamburg (Germany), and the Musee Royale de l'Afrique Centrale (Belgium) for the collection and successful transfer of pure tilapia stocks from Africa to Philippines. The GIFT partnership was also expanded to include selected tilapia farmers in Philippines who participated in on-farm experiments during the second phase of the project.

8.4 Development of Improved Nile Tilapia Strain

Several biotechnological techniques can be applied to farmed aquatic species for the genetic enhancement and management of genetic resources. However, WorldFish Center and partners were extra cautious in the manner in which the GIFT research was conducted (Acosta and Williams 2001). Emphasis was laid on choosing a breeding program that is environment friendly, sustainable, and globally acceptable. Primary consideration was made to ensure that research on GIFT would result in a technology that is compatible with the needs and capacity of the developing countries, and would give continued improvement and consequently maximum benefits (both in the short-term and long-term). Also, the technology recognizes that genetic enhancement and genetic conservation are interdependent and have to be applied in a highly responsible manner.

FAO (2001) cited the following factors that determine or influence the appropriateness of the biotechnologies for application in developing countries: their environmental impact and their impact on human health; the status with respect to intellectual property rights; the status with respect to bio-security regulations and controls; the degree of access to the biotechnologies; the level of capacity building or resources required to use them; financial costs; their ethical dimension; and their impact on food productions and food security.

It is for the abovementioned reason that the program on GIFT had chosen the traditional selective breeding as the technological approach for genetic enhancement. The development of GIFT does not involve the transfer of genes from one species to another; hence, *it is not genetically modified.*

8.4.1 Establishment of a Base Population

In view of the poor genetic status of the Asian tilapia stocks and to establish a wide genetic base before starting the genetic improvement program, wild Nile tilapia populations were brought from Ghana, Egypt, Kenya,, and Senegal during 1988–1989. Four strains of tilapia being used by commercial farmers in Asia (from Israel, Singapore, Taiwan Province of China, and Thailand) were also used in the genetic improvement program, bringing the total number of strains used in the program to eight (Eknath et al. 1993; Dey and Gupta 2000).

Performance of the pure bred and crossbred groups among the eight strains mentioned earlier, was assessed, and based on their performance, 25 best performing groups were selected to form a genetically mixed base population (synthetic population), and the selection program was initiated (Eknath 1994, 1995; Eknath and Acosta 1998). As the main problem in tilapia aquaculture in Asia was the deteriorating growth performance, selection for growth was identified as the primary trait for the genetic improvement program.

8.4.2 Selective Breeding Methodology

The strategy followed by the GIFT project for selective breeding was a combined between-family and within-family selection. The GIFT project ended in 1997, by which time, five generations of selection were undertaken. Eknath and Acosta (1998) and Eknath et al. (1998) indicated that an accumulated genetic gain of 85% over the base population with 12–17% gain per generation was achieved. ADB (2005) in its review of the GIFT project reported a genetic gain of 19.1, 13.5, 9.2, 17.8, and 6.2% during the first five generations of improvement. WorldFish Center (Khaw et al. 2008) estimated the genetic change in GIFT fish by comparing the growth performance of the progeny produced from cryopreserved spermatozoa from the base population with that produced by freshly collected spermatozoa from the ninth generation, and arrived at the estimated total genetic change in live weight, which was 64% over nine generations or 7.1% per generation (Ponzoni et al. 2008). They indicated that the reported higher genetic gain by Eknath et al. (1998) might be due to the inadvertent selection of fish to be mated as controls, and accumulation of sampling problems during the selection.

The selectively bred strain developed from the GIFT Project came to be known as the GIFT strain (Plate 8.1). Studies have proven that the GIFT fish which is genetically improved through selective breeding has a proven growth potential. Ponzoni et al. (2005) reported that despite several generations of selection undergone by the fish, the population still has additive genetic variance to enable further improvement. This finding is supported by the selection response observed (10%) after just one round of selection conducted in an experiment in Malaysia. Studies found that GIFT is a superior Nile tilapia strain, from which farmers can benefit due to its fast growth rate.

Plate 8.1 The improved strain of Nile tilapia (GIFT). (Photo credit: From the photo file of WorldFish Center)

The improvement in the latter trait was achieved without any deterioration in the survival rate, which has also remained high (Khaw et al. 2008). Other authors have also shown that GIFT and GIFT-derived strains are good genetic material for continued selective breeding. Romana-Eguia et al. (2004) has confirmed the genetic variability of GIFT, which is a good basis for selection.

8.4.3 Evaluation of GIFT Strain in Asia

The development of GIFT strain of Nile tilapia was hailed as a landmark development in the history of genetic improvement of tropical finfish. The partners of the project recognized that the task would be incomplete unless the strain was translocated in a responsible manner and in a way that this would bring about positive socioeconomic change and negligible impacts on the ecosystems. In this connection, the GIFT program held an international expert consultation (International Concerns in the Use of Aquatic Germplasm) in 1992 to discuss the strategies and safeguard measures for the transfer and distribution of GIFT germplasm (Eknath 1995). Following the recommendations from this meeting, the second and fourth generations of GIFT strain were evaluated in different farming systems and agro-ecological zones in Bangladesh, China, Philippines, Thailand, and Vietnam for their genetic and economic performance.

Experimental testing was conducted for GIFT fish together with the "best" available strains in the participating countries. The local existing strains used were the Thai strain and red tilapia strain in Bangladesh; Nile tilapia Egypt strain, "78" strain, "88" strain, and American strain in China; Chitralada strain (Chitralada I strain in 1995), sex-reversed GIFT (SRT GIFT), sex-reversed Chitralada I (SRT Chitralada I), and genetically male tilapia (GMT) strains in 1996, in Thailand; and Egypt strain, Thai strain, and Vietnam strain in Vietnam. In Philippines, on-station experiments were carried out under the GIFT Project; hence, only on-farm trials were conducted (Dey et al. 2000; ADB 2005).

In all the participating countries, the GIFT strain performed much better than the best non-GIFT local strain. On an average farm, the GIFT strain gave 18% higher body weight at harvest in China to 58% in Bangladesh. In Philippines, the synthetic base population developed through cross breeding of various populations used in the breeding program showed 60% increased growth when compared with the Philippine-farmed strains. Together with the subsequent responses to selection, this represents a claimed growth performance advantage of about 125% for GIFT over Philippine-farmed strains. In other words, in comparable environments, GIFT should reach a harvest size in less than half the time taken by unimproved tilapia strains (ADB 2005).

It was also demonstrated that GIFT strain is a neutral technology with respect to feed and fertilizer use, which means that small farmers who use less feed and fertilizers and big farmers who use more feed and fertilizers will get commensurate benefits (Dey and Eknath 1997; ICLARM 1998; Dey et al. 2000; Dey and Gupta 2000).

8.5 Dissemination of Improved Tilapia Strain

8.5.1 International Program for Responsible Transfer of Improved Genetic Material

Genetically improved fish has been targeted as a source of potential socioeconomic benefits. However, there might be risks associated with the intentional and nonintentional introduction of this fish. In the case of GIFT strain, the risks associated with its release were reported to be very low, with considerable benefits in using this strain in places where alien populations of tilapia had already become established. The widespread dissemination of the improved strain has also raised concerns over the capacity of the many countries in terms of policies and strategies for dissemination, and to assess and manage biosafety risks associated with trans-boundary movements (WorldFish Center 2002; Kapuscinski et al. 2007; NACA 2007). Accordingly, the transfer of GIFT fish was made under highly precautionary policies of the WorldFish Center and in accordance with the internationally accepted standards of FAO, the Convention on Biological Diversity (CBD), and other international bodies concerned with legal, political, ethical, and economic issues for conservation and sustainable use of aquatic genetic resources (Gupta and Acosta 2004).

The International Network of Genetics in Aquaculture (INGA) established by the WorldFish Center and its partners in 1993 provided the international partnerships and links for the dissemination and further development of GIFT (Gupta and Acosta 2001a, b). INGA assisted its members and other countries in the sharing of GIFT for the development of national breeding programs and dissemination to farmers. Exchanges of GIFT and other improved germplasm were made following the protocols and quarantine procedures (Box 8.1) based on international codes of practice. Material transfer agreements formulated by the network were/are being used as guidelines by the countries or organizations acquiring the GIFT fish and other germplasm through a network (Gupta and Acosta 2001b). The GIFT strain has been disseminated to Bangladesh, Côte d'Ivoire, Egypt, the Fiji Islands, India, Indonesia, Kenya, Lao People's Democratic Republic, Malaysia, Papua New Guinea, People's Republic of China, Thailand, and Viet Nam.

8.5.2 Commercial Dissemination of GIFT-Derived Strain

In 1999, the fifth generation of GIFT strain was transferred to GenoMar ASA (formerly BioSoft AS) by the GIFT Foundation International Inc., under an agreement. Using the trademark name, GenoMar Supreme Tilapia (GST), the new strain has been further improved by selective breeding and by introducing DNA typing as an identification system, replacing the conventional physical tags (GenoMar 2008). The GST is being disseminated commercially through GenoMar's own hatcheries in Philippines and China. Dissemination is also being made through its partner hatcheries in three continents (Asia, Africa, South America). In Africa, GenoMar has already supplied the GST to partner hatcheries in Zambia and Angola, and is currently carrying out a feasibility study on establishing a partnership in Uganda (GenoMar 2008).

In terms of performance, GenoMar (2008) reported that by applying state of the art breeding technology, it has increased the genetic gain per year by more than 35%, when compared with conventional breeding schemes. GenoMar now produces a new generation every nine months and the annual genetic gain is estimated to be more than 15% of the growth. In China, experiments showed that GST grew more than twice as fast as the local commercial strain. GenoMar ASA company has established a large hatchery (GenoMar Supreme Hatchery China) in Hainan province for a large-scale dissemination of GIFT strain to the world's largest tilapia market (Zimmerman 2002, cited by Lai and Yang 2008).

8.5.3 Development of Strategy for Effective Dissemination and Maintenance of GIFT Strain

Most of the INGA member counties have developed programs for further improvement and dissemination of GIFT tilapia. However, these programs have not been very effective. In response to this and to determine the kind of support needed by the member countries, WorldFish Center convened the international "Workshop on Dissemination of Improved Fish Strains: Country-Specific Action Plans" on September 21–22, 2005 in Shanghai, China, to discuss ways to strengthen the ongoing genetic improvement and dissemination programs of the member countries (Ponzoni et al. 2006). As a follow-up to this activity, WorldFish Center conducted an inventory of GIFT strain in all the countries where it is present. This activity is part of a broader project aimed at establishing a formal inventory of GIFT stock, strengthening the GIFT breeding program, and implementing a more effective multiplication and dissemination strategies of improved fish. This will ensure that the greater productivity of GIFT is maintained and further enhanced, and that the benefits are captured by fish farmers (WorldFish Center 2005). WorldFish Center (2004) has also published a manual on GIFT technology that is useful for those who are involved in selective breeding and those who wish to initiate selective breeding programs.

8.5.4 Introductions of GIFT Fish from Asia to Africa

It has been recognized that introductions of the GIFT fish to areas where tilapia are endemic (e.g. in parts of Africa) may pose substantially high risk and could endanger the native tilapia genetic resources. Accordingly, WorldFish Center at the commencement of the GIFT Project developed a policy not to introduce the GIFT strain into Africa (Pullin 1994; WorldFish Center 2007; Anonymous 2008). Recently, in view of the scarcity of examples of environmental damage from tilapia, the increasing interest from African farmers on GIFT fish and the realization that introducing GIFT-selective breeding as a technology into Africa is only a few years different from the actual introduction of the GIFT fish, WorldFish Center has changed its position on re-introduction of GIFT fish to Africa (FAO 2006b). WorldFish Center has decided to make the GIFT fish available to the African Government on the condition that such request is made by a relevant government authority, with a well-defined strategy to maintain and disseminate GIFT, and a clear plan for the management of environmental and biodiversity risks (WorldFish Center 2007).

The CBD, an international agreement with 182 member countries including 53 in Africa, requires parties to prevent the introduction of, control, or eradicate those alien species that threaten the ecosystems, habitats, or species (Hill and Sendashonga 2004). Nile tilapia is not a native to all African countries; hence, introduction of GIFT fish to Africa has to be based on individual cases. The decision to introduce the GIFT fish to Africa will still depend on whether the state could comply with the CBD's requirements, while at the same time address the development objective (i.e. increase the availability of fish protein for the poor by introducing and farming the improved Nile tilapia strain). It also depends on what is considered a priority by the state or country that will receive the GIFT fish (i.e. development objective vs. conservation objective).

Irrespective of the decision that one takes, there will be a trade-off. For instance, if the former decision is taken, then there is a risk that the GIFT stocks will escape into the natural environment and will introgress with the wild Nile tilapia stocks, if it is endemic to that country. Even if GIFT fish is not introduced into Africa and the countries decide to undertake genetic improvement of their native stocks, it would still lead to changes in composition of alleles which will not be much different from introducing the GIFT fish. If the GIFT fish is not transferred to the African countries which are the original source of founder stocks used in developing the GIFT strain, they will be deprived of the benefits from the utilization of the improved strain. Moreover, in view of the emergence of the more enterprising African farmers and growing interest in aquaculture, there is a greater risk of not introducing the GIFT fish, as otherwise, they might introduce other fish that might do more harm to biodiversity and environment. In such instances, there may be more environmental risks, as there is no assurance that control and monitoring mechanisms plus safeguard measures will be implemented.

8.5.5 Public–Private Partnerships

In view of the progress made on genetic improvement, advancements in farming technology, and increased domestic and global demand for the fast-growing strains of fish, there has been an increasing participation of the private sector as multipliers or for the production and dissemination of improved tilapia strains (including the GIFT strain) since the late 1990s (Acosta et al. 2004). For instance, in Philippines, where the GIFT strain was disseminated primarily through the GIFT Foundation's licensing arrangements with privately owned hatcheries in the country, it was reported that over 200 million fingerlings had been disseminated for the first 5 years of its operation. In Thailand, GIFT accounted for 46% of the 576 million tilapia seed sold by these hatcheries in 2003 (Pongthana 2004).

8.6 Creating an Environment for Sustainability of GIFT

8.6.1 Capacity Building in Developing Countries

The major objective of the GIFT Project was to develop methods for genetic improvement of tropical fish and disseminate the techniques developed for genetic enhancement of other species (e.g. carps) in different countries. These outputs, particularly, the techniques developed from the project, have been and are being disseminated and adopted by national programs to help improve food security in developing countries. To address this and to sustain the genetic gains achieved from the GIFT project, WorldFish Center and its partners have provided greater focus on capacity building and formulating genetic improvement programs.

The INGA played a crucial role in strengthening the capacity of the developing countries in applied breeding and genetics. Based on the training needs assessment of

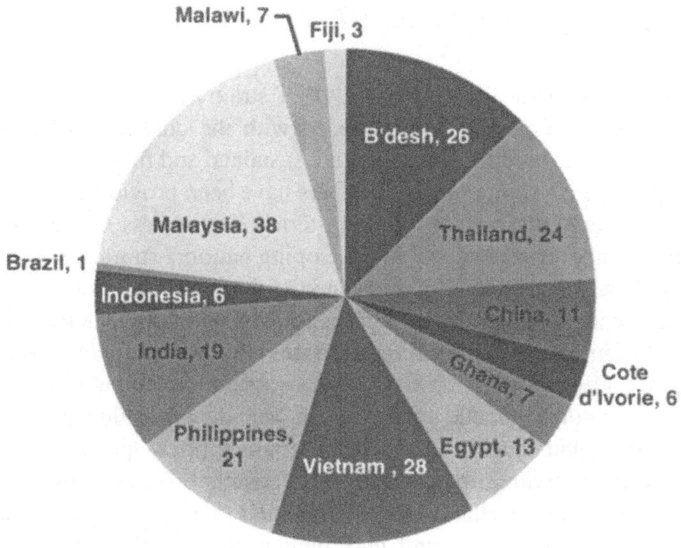

Fig. 8.2 Number of technical persons trained per country under International Network on Genetics in Aquaculture (INGA)

the member countries, INGA, through the WorldFish Center's genetic improvement programs, organized series of capacity building programs that focused on quantitative genetics and applied breeding. As of 2004, a total of 210 participants from 14 countries of Asia, Africa, and Latin America had been trained (Fig. 8.2).

The strengths of INGA lie primarily on commitment and cooperation of the network members, the participatory mode of operations, and the complementarities of skills among its members. However, limited financial and human resources have constrained the continuing operations of INGA. Financial assistance from the Government of Norway ended in 2003. Since then, in view of the shifting priorities of the traditional donors, sourcing funds to sustain the activities of a research network such as the INGA has become a major challenge. WorldFish Center recognized this challenge; hence, to ensure that member countries are continuously assisted in their breeding programs, it explored ways on how the activities related to training could be built into its various donor-funded projects. For instance, through a project funded by UNDP, WorldFish Center has strengthened the capacity of Egypt, Cote d'Ivoire, Ghana, and Malawi in aquaculture genetics and has assisted them in developing their national breeding programs (WorldFish Center 2006). As part of this activity, a user manual describing in details the key operations in a selective breeding program for tilapias was published (WorldFish Center 2004). This manual is a follow-up to the compilation of the GIFT protocols prepared and disseminated in member countries of INGA when the GIFT project ended (Acosta and Eknath 1998).

8.6.2 Follow-Up Program on GIFT

With the increase in the availability of the GIFT strain and other new improved fish breeds, developing countries are now faced with the challenge of ensuring that genetic gains from the breeding programs are sustained and further enhanced. In this connection, WorldFish Center and its partners have been providing greater focus on developing a strategy that will ensure long-term effectiveness of the breeding programs in the member countries and in developing national strategies, guidelines for the effective dissemination, and maintenance of the improved strains.

The draft of country-specific action plans for genetic improvement of the selected species and dissemination of the improved strains have now been developed (Ponzoni et al. 2006; WorldFish Center 2006). The plans when implemented are expected to strengthen the ongoing genetic improvement and dissemination programs of the member countries, and develop additional programs for new species.

At present, the breeding objective for GIFT fish is broadened by considering a number of additional important traits, such as filet yield, flesh quality, sex ratio in response to thermal treatment, and may include disease resistance in the future (Nguyen and Ponzoni 2006).

8.7 Development of Program to Ensure Maximum Benefits from GIFT

The GIFT project is a good example of how multidisciplinary and interactive research initiatives with successful partnerships could progress into a global program for genetic improvement of tropical finfish. Figure 8.1 illustrates the stages of globalization of methods and outputs from the project.

Research efforts of WorldFish Center and its partners from Philippines and Norway led to the development of two major outputs of genetics research – GIFT strain and the method for genetic enhancement of tropical finfish. Through various programs, GIFT research outputs are now being disseminated widely and are already showing positive impacts in terms of improving fish productivity and economic benefits to the developing countries.

8.7.1 Impacts of GIFT Technology

8.7.1.1 Application of GIFT Technology in Carps and Other Species in Asia

In view of the potential benefits of selective breeding, the GIFT technology has been applied for genetic improvement of various carp species in six Asian countries, viz., silver barb (*Barbodes gonionotus*) in Bangladesh and Thailand, rohu

(*Labeo rohita*) in India and Bangladesh, common carp (*Cyprinus carpio*) and mrigal (*Cirrhinus mrigala*) in Vietnam, and blunt snout bream (*Megalobrama amblycephala*) in China (Fig. 8.3).

In Bangladesh, encouraging results were obtained from genetic improvement of silver barb, *B. gonionotus* through several generations of selection (up to the fourth generation). The selected strain was found to be 28% superior to nonselected existing strain, indicating that the development of "super strain" of silver barb is possible (Hussain et al. 2002).

In China, the selection experiments initiated with blunt snout bream (*M. amblycephala*) indicated that the fifth generation of selected strain increased 30% relative to the control group. The results demonstrate that positive selection is a powerful tool to improve the economic traits of blunt snout bream. Hence, in 2000, the sixth generation of selected bream was certified by the Chinese Ministry of Agriculture as a good breed for aquaculture (Li 2002).

Success was reported on selective breeding of rohu (*Labeo rohita*) in India. The improved rohu (Jayanti strain) was developed from a base population composed of five wild riverine stocks and one farm stock. This strain which is considered as the first genetically improved fish of India has already been distributed to several local hatchery owners to provide better quality seed to fish farmers (Das Mahapatra et al. 2006).

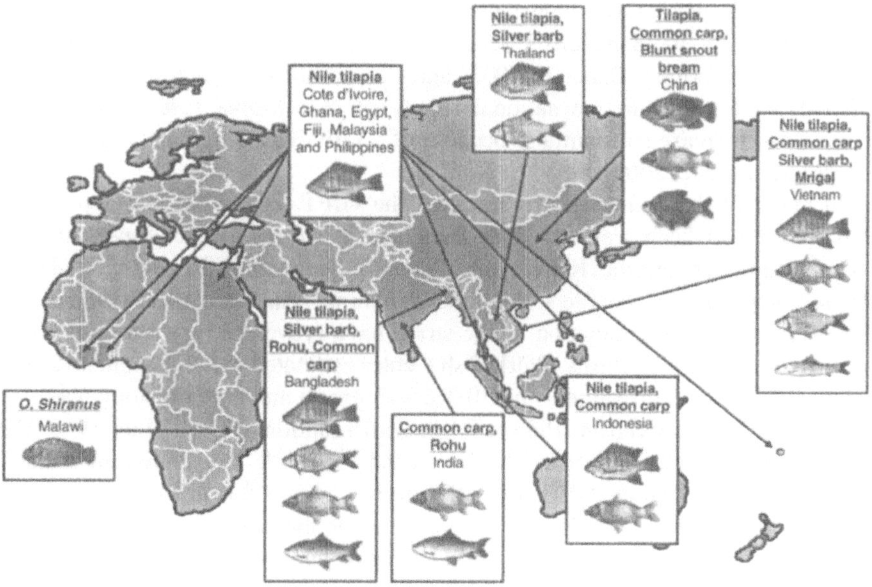

Fig. 8.3 National breeding programs in the member countries of INGA with focus on carps and tilapias

In Thailand, a selection program for increased growth rate in Chao Phraya strain of silver barb, *B. gonionotus* (Chao Phraya strain) gave encouraging results. After five generations of selection, the selected line had significantly greater body weight than the control (Pongthana et al. 2006).

In Vietnam, a selection program was conducted for the common carp (*C. carpio*) to develop better performing strains used for aquaculture. Growth evaluation trial indicated that improved common carp was 50% bigger when compared with the local common carps, while the survival rate was comparable (90%). A follow-up experiment was conducted at several farms in Hanoi, Vinh Phuc, Bac Giang, and Bac Can provinces, and they found that the common carp improved by selective breeding grew 30% faster than the local common carps (Dan and Thien 2002).

8.7.1.2 Impacts of GIFT Strain

Through exchanges of germplasm which enabled the developing countries to have access to GIFT strain, most member countries of INGA are able to initiate their breeding programs. Now, most of these countries have improved strains (GIFT and GIFT-derived) that are being disseminated to both public and private farms, and comprise a substantial proportion of their overall tilapia production.

The substantial impacts of GIFT and GIFT-derived strains on farmed tilapia production are evident from the increasing shares in tilapia seed supply. In Philippines, the survey commissioned by the Asian Development Bank found that GIFT and GIFT-derived strains accounted for 68% of the total tilapia seed produced in 2003. In Thailand, GIFT contributed to 46% of all national tilapia seed production in 2003. In Vietnam, GIFT seed contributed an estimated 17% to national production of farmed tilapia in 2003, and the overall contribution of GIFT and GIFT-derived strains to the national supply of Nile tilapia seed in Vietnam is expected to increase substantially because of a GIFT-based national tilapia breeding program (ADB 2005).

In many of the countries that received the GIFT fish (GIFT developed from Philippines had been transferred to Bangladesh, Côte d'Ivoire, Egypt, the Fiji Islands, India, Indonesia, Kenya, Lao People's Democratic Republic, Malaysia, Papua, New Guinea, People's Republic of China, Thailand, and Vietnam), the development and dissemination of the strain have proven to be successful investments with economic returns. WorldFish Center estimated that the economic internal rate of return on investments in GIFT development and dissemination was more than 70% over a period from 1988 to 2010, with adoption of GIFT since 1996 (Deb and Dey 2004 cited by WorldFish Center and PRIMEX Inc. Philippines. 2007). There is evidence that the GIFT and GIFT-derived strains that are already available in many developing countries are responsible for the present increase in tilapia production from a wide range of farming systems and fish supply to a wide range of consumers, including the poor (ADB 2008). In many of these countries, GIFT tilapia generally performed better than the existing farmed tilapia. For instance, in Thailand, it was found that the average percentage filet yield from GIFT (38%) is higher than that of Chitralada Nile tilapia (34%) which was formerly the strain most widely farmed in the country (Rutten et al. 2002, 2004). In China, results of a 2-year experiment showed that GIFT fish was better performing in the following aspects: (1) 5–30% faster in growth rate; (2) 20–30% higher in yields per area; (3) 2–3 times higher in seinability; and (4) 60–110% higher in salinity tolerance. In view of these, the GIFT fish has been certified by the China's Ministry of Agriculture as a good strain for dissemination in the country (Li 2002).

The introduction and dissemination of GIFT have generated significant rural incomes and employment, and contributed to human nutrition, especially among the poor, as tilapia is a relatively low-priced fish. Tilapia farming provides an attractive livelihood for hatchery operators and fish farmers. The contribution of GIFT to employment generation is significant. A study undertaken by ADB (2005) on tilapia hatcheries and growout ponds in Philippines and Thailand indicated net returns of about $5,000/ha/year in the case of hatcheries, and $1,783–$4,241/ha/crop cycle (of 4–8 months) in the case of growout operations (excluding the fish consumed by households on farm). In Philippines, the 4-month crop cycles, allowing two crops a year to fit the climatic and seasonal conditions, are common. In some parts of the central and Northern Thailand, tilapia farmers follow 8-month crop cycles once a year and 6-month crop cycles twice a year.

For GIFT in growout cages, indicative net returns ranged from $103 to $390/cage/cycle, depending on the cage size and crop cycle duration, among other factors (Table 8.1). At Lake Taal, Philippines, a fish farmer with four cages who harvests tilapia twice a year can earn net returns of $3,120 a year. In Northern Thailand, a

Table 8.1 Examples of indicative net returns from farming Genetic Improvement of Farmed Tilapias (GIFT) in 2003 (source: ADB 2005)

	Philippines		Thailand	
Source and unit of measure	In Pesos (P)	In US Dollars[a]	In Baht (B)	In US Dollars[b]
Hatchery (per ha/year)	275,000	5,074	200,000	4,819
Growout pond (per ha/crop)				
4-month crop cycle	101,188	1,867	–	–
6–6.5-month crop cycle	–	–	74,000–137,000	1,783–3,301
8-month crop cycle	–	–	176,000	4,241
Growout cage (per cage/crop)[c]				
Cage size 10×10×10 m (5–6-month crop cycle)	21,119	390	–	–
Cage size 3×6×2.5 m (4-month crop cycle)	–	–	4,285	103

[a]$1 = P54.2
[b]$1 = B41.5
[c]Cage sizes and stocking rates vary, and the total household incomes depend on the number of cages operated
Sources: Key informant interviews and surveys

farmer who farms GIFT in cages along the Ping River as a secondary income source has earned net returns of $1,236 a year (4 cages, 3 crop cycles a year) (ADB 2005).

The study further indicated that dissemination and adoption of GIFT and GIFT-derived strains contribute significantly to the generation of employment in tilapia farming. For example, at least 280,000 people in Philippines, 200,000 people in Thailand, inclusive of their families, directly and indirectly benefit annually from employment generated by tilapia farming alone (ADB 2005). The poor and small-scale farmers are among those who benefit from employment in tilapia farming and its associated activities. Based on the hatchery surveys conducted for this study, GIFT and GIFT-derived hatcheries in Philippines and Thailand generated about 68 and 45% employment, respectively, of their national tilapia hatchery workforces in 2003.

Though GIFT, as indicated earlier, is not a genetically modified tilapia, there were concerns that their farming could result in adverse impacts on the environment and biodiversity in areas/countries where it is introduced or will be introduced. However, some of the reviews undertaken (Bentsen et al. 1992; De Silva 1997) have not indicated that development and dissemination of GIFT is unlikely to cause serious impacts either on the environment or biodiversity This view has been further strengthened by a global review of the environmental impacts of tilapia, that Nile tilapia, although potentially invasive as an alien species, has had far fewer adverse environmental impacts than other tilapia species introduced and disseminated for aquaculture (Pullin et al. 1997 as quoted by ADB 2005).

The ADB study undertaken in Philippines, Thailand, and Vietnam has demonstrated the economic benefits that could be obtained from utilization of the GIFT and GIFT-derived strains. However, information is still lacking regarding the ex-post impact of GIFT technology in Asia. To address this, in 2006, WorldFish Center initiated a BMZ-funded research project that would assess the impact of GIFT introduction on small-scale farmers. The project study is expected to demonstrate the ex-post impact of the technology on aquaculture productivity, farm income, and effects on poverty reduction (Pemsl 2005).

8.8 Lessons Learned

- Partnerships among the national institutions in developing countries, advanced scientific institutions, and regional and international organization can play a major role in accelerating the research and the success of research and development, as has been demonstrated. Partnerships between developing and developed country institutions has resulted in the transfer of technology and collaborative research projects. This has also resulted in strengthening of the capacity of developing country institutions. INGA and programs of WorldFish Center have provided a mechanism for strengthening the capacity of the member countries in managing the GIFT which is vital to ensure sustainability of the improved strain.

- As has been demonstrated in Philippines and Thailand, partnerships between the government institutions and the private-sector hatchery industry has resulted in faster and wider dissemination of the GIFT seed to the farmers.
- Experience from the Asian and African member countries of INGA has indicated that success of genetic research and development depends on: (a) strong aquaculture sector in the country and demand for improved seed; (b) networks for dissemination of seed; (c) capabilities and capacity of national institutions (human, financial, and infrastructure resources) in fish genetics research; (d) long-term commitment of government resources; and (e) supportive policies for conservation of biodiversity and environment.
- The GIFT experience highlighted the need for institutionalization and implementation of a country-specific follow-up program once improved strains have been developed and disseminated. It is essential that links and coordination with the main breeding nucleus are maintained and strengthened so that application of effective management and dissemination strategies to these programs can be facilitated (Ponzoni et al. 2006). While control of a central genetic nucleus remains in the public sector, a dedicated competent human support service has to be made available to oversee and instigate the gradual involvement of the private sector in the program for reproduction and commercial dissemination of the improved strain (Ponzoni, personal communication, cited by Acosta et al. 2006).
- Public and private sectors have an important role to play in the development and effective dissemination of improved fish seed. The two sectors need to complement each other in developing an efficient seed supply chain to allow the farmers gain access to high-quality seed and thus, contribute to sustainable aquaculture. There is a need for country-specific policies that will harmonize their roles and activities in the overall program of genetic improvement and dissemination, and improve the accessibility of poor farmers to improved fish. Complementary and supportive policies need to be formulated to facilitate effective cooperation arrangements among the sectors engaged in dissemination of improved fish. The GIFT experience has revealed that while participation of the private sector is important to facilitate wider dissemination of the improved strain, and the governments have a crucial role to play in ensuring healthy competition between the public and private sectors. Apart from the private sector, the other key players that can be tapped to deliver the services that public sector may fail to provide are the nongovernment organizations.
- As has been revealed by some of the studies undertaken, the development and dissemination of GIFT, GIFT-derived, and other Nile tilapia strains have not caused any significant adverse impacts on the existing aquaculture or on the natural environment and biodiversity in the Asia-Pacific region. However, taking into consideration that Asia-Pacific region has a wealth of freshwater biodiversity and habitats, areas containing this natural heritage should, wherever possible, be kept off limits to aquaculture, whether of native species or aliens, including tilapia. Such areas would contain the wild genetic resources for future breeding programs of Asian-farmed fish, and would serve as in situ gene banks for this purpose, in addition to their conservation and amenity values.

Box 8.1 A country planning to import new or exotic species has to sign a Material Transfer Agreement which states that the recipient should agrees to

- Abide by the provisions of the Convention on Biological Diversity
- Preclude further distribution of germplasm to locations where it could have adverse environmental impact
- Not Neither claim ownership over the material received, nor seek intellectual property rights over the germplasm or related information
- Ensure that any subsequent person or institution to whom they make samples of germplasm available is bound by the same provision
- The responsibility to comply with the country's biosafety, and import regulations and any of the recipient country's rules governing the release of the genetic material that is entirely theirs
- Follow with the quarantine protocols suggested by ICLARM/INGA, and abide by the International Codes of Transfer of Germplasm, if the germplasm is transferred beyond the boundaries of their country

References

Acosta, B.O., and A.E. Eknath, eds. 1998. *Manual on genetic improvement of farmed tilapia (GIFT) research methodologies*, 250. Makati City: ICLARM.

Acosta, B.O., and M. J. Williams. 2001. The role of an international research organization in tilapia aquaculture. In *Tilapia 2001: Production, marketing and technological development*, ed. S. Subasinghe and T. Singh, 49–64. Proceedings of the International Trade Conference on Tilapia, 28–30 May 2001. Kuala Lumpur.

Acosta, B.O., R.C. Sevilleja, B.M. Rodriguez Jr., T. Abella, M. Tayamen, and M.V. Gupta. 2004. Paper presented at the International Development Research Centre of Canada/Research on Knowledge Systems (IDRC/RoKS) Roundtable Meeting held in Ottawa, Canada, 26–29 April 2004

Acosta, B.O., R.C. Sevilleja, M.V. Gupta, B.M. Rodriguez, Jr., T. Abella, and M. Tayamen. 2006. Public and private partnerships in tilapia research and development: An overview of Philippine experience. In *Public and private partnerships in aquaculture: A case study on tilapia research and development*, eds. B.O. Acosta, R.C. Sevilleja and M.V. Gupta, 1–9. WorldFish Center Conference Proceedings 72. Penang.

ADB. 2005. *Impact evaluation study on the development of genetically improved farmed tilapia and their dissemination in selected countries*, 77. Manila: Asian Development Bank.

Anonymous. 2008. Available at http://www.worldfishcenter.org/.

Bentsen, H., T.Berg, and J. Johan Schei. 1992. *Environmental effects of release and dissemination of improved Nile tilapia*, 10. Report prepared by the Agricultural University of Norway for the United Nations Development Programme, Division for Global and Interregional Programmes.

Beveridge, M., and D. Baird. 2000. Diet, feeding and physiology. In *Tilapias: Biology and exploitation*, eds. M. Beveridge, and B. McAndrew, 59–87. Dordrecht: Kluwer Academic Publishers.

Dan, N.C. and T.M. Thien. 2002. Status of national breeding program in Northern Vietnam. Paper presented at the International Network on Genetics in Aquaculture Expert Consultation on Development of Strategies and Plans for Dissemination of Improved Fish Breeds to Small-scale Farmers. Pathumthani, Thailand, 4–7 June 2001.

Das Mahapatra, K., R.K. Jana, J.N. Saha, B. Gjerde, and N. Sarangi. 2006. Lessons from breeding program of rohu. In *Development of aquatic animal genetic improvement and dissemination programs: Current status and action plans*, eds. R.W. Ponzoni, B.O. Acosta, and A.G. Ponniah, 34–40. WorldFish Center Conference Proceedings 73, 120.

Deb, U. and M. Dey. 2004. The history and impacts of GIFT and Dissemination and Evaluation of Genetically Improved Tilapia. Penang: WorldFish Center.

Delgado, C.L., N. Wada, M.W. Rosegrant, S. Meijer, and M. Ahmed. 2003. *Fish to 2020: Supply and demand in changing global markets*. WorldFish Center Technical Report, 62. Washington, DC: International Food Policy Research Institute.

Dey, M.M. and A.E. Eknath. 1997. Current trends in the Asian tilapia industry and the significance of genetically improved tilapia breeds. INFOFISH: Kuala Lumpur.

Dey, M.M. and M.V. Gupta. 2000. Socioeconomics of disseminating genetically improved Nile tilapia in Asia: An introduction. *Aquaculture Economics and Management* 4: 5–11.

Dey, M.M., A.E. Eknath, L. Sifa, M.G. Hussain, T.M. Thien, N.V. Hao, S. Aypa, and N. Pongthana. 2000. Performance and nature of genetically improved farmed tilapia: A bioeconomic analysis. *Aquaculture Economics and Management* 4: 85–108.

De Silva, S. 1997. Popularization of Genetically Improved *Oreochromis niloticus* (GIFT fish) in Asia: Environmental considerations. Paper presented at the Final Workshop of the Dissemination and Evaluation of Genetically Improved Tilapia Species in Asia (DEGITA) Project, 18–19 February 1997, ICLARM, Makati City, Philippines.

Eknath, A.E. 1994. The project of genetic improvement of farmed tilapia (GIFT) from modest beginning to an International Network on Genetics in Aquaculture. In *Tilapia farming, genetic improvement and advances on culture technology*, eds. I.G. Villacorta and L.A. Dureza, 9–13. Proceedings of the Third National Symposium and Workshop on tilapia farming. Los Banos: PCAMRD book series, no. 18/1994.

Eknath, A.E. 1995. Managing aquatic genetic resources. Management example 4. The Nile tilapia. In *Conservation of fish and shellfish resources: Managing diversity*, eds. J.E. Thorpe, G. Gall, J.E. Lannan, and C.E. Nash, 176–194. London: Academic Press, Harcourt Brace Company Publishers.

Eknath, A.E. and B. Acosta. 1998. Genetic improvement of farmed tilapias (GIFT) project Final Report, March 1998 to December 1997. ICLARM: Makati City, Philippines.

Eknath, A.E., J.M. Macaranas, L.Q. Augustin, R.R. Velasco, M.C.A. Ablan, M.J.R. Pante, and R.S.V. Pullin. 1991. Biochemical and morphometric approaches to characterise farmed tilapias. *Naga, The ICLARM Quarterly* 14: 7–9.

Eknath, A.E., M.M. Tayamen, M.S. Palada-de Vera, J.C. Danting, R.A. Reyes, E.E. Dionisio, J.B. Gjedrem, and R.S.V. Pullin. 1993. Genetic improvement of farmed tilapias: The growth performance of eight strains of *Oreochromis niloticus* tested in eleven different environments. *Aquaculture* 111: 171–188.

Eknath, A.E., M.M. Dey, M. Rye, B. Gjerde, T.A. Abella, R. Sevilleja, M.M. Tayamen, R.A. Reyes, and H.B. Bentsen. 1998. *Selective breeding of Nile tilapia for Asia*, 89–96. Proceedings of the 6th World Congress on genetics applied to livestock production. Armidale, Australia: University of New England, 27.

FAO. 2001. *Research and development of appropriate genetic biotechnologies for the fishery sector in developing countries*, 44. Report of the Third Session of the Advisory Committee on Fisheries Research. FAO Fisheries Report No. 639. Rome: FAO.

FAO. 2006a. *World review of fisheries and aquaculture*. Available at: ftp://ftp.fao.org/docrep/fao/009/a0699e/10699e01.pdf.

FAO. 2006b. *Species choice in aquaculture; domestication, processes, genetic improvement and their role in sustainable aquaculture*, 8. Report of the FAO Advisory Committee on Fisheries Research. Rome, Italy. 17–20 October 2006.

FAO. 2007a. *The next frontier in genetic research: Under water*. FAO NewsRoom. Available at: http://www.fao.org/newsroom/en/news/2007/1000593/index.html.

FAO. 2007b. *Concern over situation of high seas fish species. Strengthening fisheries management in international waters "a major challenge" – FAO report*. Available at: http://www.fao.org/newsroom/en/news/2007/1000505/index.html/.

Fitzsimmons, K. 2008. *Review – global advances in tilapia production and marketing*. Available at: http://az/arizona.edu.azaqua/ista/Busan 2008ppt.

GenoMar. 2008. *Tilapia breeding*. Available at: http://www.genomar.no/?did=9073496&aid=9073399.

Gupta, M.V., and B.O. Acosta. 2001a. *Development of global partnerships for fish genetics research – a success story*. Paper presented at the technical workshop on methodologies, organization and management of global partnership programmes, 9–10 October 2001, Rome, Italy.

Gupta, M.V., and B.O. Acosta, 2001b. Networking in aquaculture genetics research. In *Fish genetics research in member countries and institutions at the International Network on Genetics in Aquaculture*, eds. M.V. Gupta and B.O. Acosta, 1–5. ICLARM Conference Proceedings, 64. Manila: International Centre for Living Aquatic Resources Management (ICLARM).

Gupta, M.V., and B.O. Acosta. 2004. From drawing board to dining table: The success story of GIFT project. *NAGA. WorldFish Center Quarterly*, 27: 4–14. (July–December 2004).

Hill, R., and C. Sendashonga. 2004. The role of the convention on biological diversity and the Cartagena protocol on biosafety in minimizing adverse effects of invasive alien species and living modified organisms. In *Use of Genetically Improved Alien Species for Aquaculture and Conservation of Aquatic Biodiversity in Africa*, eds. M.V. Gupta, D. Bartley and B.O. Acosta, 113. Penang: WorldFish Center.

Hussain, M.G., M.S. Islam, M.A. Hossain, M.I. Wahid, M.M. Dey, and M.A. Mazid. 2002. Stock improvement of silver barb (*Barbodes gonionotus* Bleeker) through several generations of genetic selection. *Aquaculture* 204: 469–480.

ICLARM. 1998. *Dissemination and evaluation of genetically improved species in Asia: Final report*. Asian Development Bank regional technical assistance no. 5558. Manila: ICLARM.

Kapuscinski, A.R., G. Dana, K.R. Hayes, S. Li, K.C. Nelson, Y.K. Nam, Z, Gong, R.H. Devlin, G.C. Mair, and W. Senanan. 2007. Risk assessment of transgenic fish: Synthesis and conclusions. In *Environmental risk assessment of genetically modified organisms, vol. 3 – Methodologies for transgenic fish*, eds. A.R. Kapuscinski, K.R. Hayes, S. Li, and G. Dana, 272–289. Washington, DC: Published by CABI International, 295

Khaw, H.L., R.W. Ponzoni, and M.J.C. Danting. 2008. Estimation of genetic change in the GIFT strain of Nile tilapia (*Oreochromis niloticus*) by comparing contemporary progeny produced by males born in 1991 or in 2003. *Aquaculture* 275: 64–69.

Lai, Q., and Y. Yang. 2008. *Tilapia culture in Mainland China*. Available at: http://ag.arizona.edu/azagua/ista/ista6/ista6web/pdf/018.pdf.

Li, S. 2002. Strategies, plans for dissemination of improved fish breeds in China. In *WorldFish Center*. Report of INGA expert consultation on development of strategies and plans for dissemination of improved fish breeds to small-scale farmers. 4–7 June 2002, Pathumthani: NAGRI.

Macaranas, J.M., R.S.V. Pullin, M.J.R. Pante, and J.B. Capilli. 1984. Genetic contamination of cultured Nile tilapia (*Oreochromis niloticus*) in the Philippines. *Kalikasan, Philippine Journal of Biology* 14: 1–8.

Macaranas, J.M., N. Macaranas, J.M. Taniguchi, M.J.R. Pante, J.B. Capilli, and R.S.V. Pullin. 1986. Elecrophoretic evidenceof extensive hybrid gene introgression into commercial *Oreochromis niloticus* (L.) stocks in the Philippines. *Aquaculture and Fisheries Management* 17: 248–258.

NACA. 2007. Report on the workshop on resarch needs in sustaining the aquaculture sector in Asia-Pacific to year 2025 and beyond. June 4–7, 2007. Rayong, Thailand, 162.

Nguyen, N.H., and R.W. Ponzoni. 2006. Perspectives from Agriculture: Advances in Livestock Breeding – Implications for Aquaculture Genetics. *Naga WorldFish Center Quarterly*, vol. 29, Nos. 3 and 4. July–December 2006.

Pemsl, D.E. 2005. *Proposal: Assessing the farm level impact of genetically improved farmed tilapia (GIFT) in China and the Philippines*, 10. Penang: WorldFish Center.

Pongthana, N. 2004. Status of dissemination of improved fish breeds in Thailand, 8. Consultancy report submitted to WorldFish Center. Penang, Malaysia.

Pongthana, N., N.H. Nguyen, and R.W. Ponzoni. 2006. *Mass selection for increased body weight at harvest over five generations in Thai silver barb (Puntius gonionotus)*. Proceedings of the 8th World Congress on Genetics Applied to Livestock Production, 13–18 August 2006, Belo Horizonte, MG, Brazil.

Ponzoni, R.W., A. Hamzah, S. Tan, and N. Kamaruzzaman. 2005. Genetic parameters and response to selection for live weight in the GIFT strain of Nile tilapia (*Oreochromis niloticus*). *Aquaculture* 247: 203–210.

Ponzoni, R.W., B.O. Acosta, and A.G. Ponniah, eds. 2006. *Development of Aquatic Animal Genetic Improvement and Dissemination Programs: Current Status and Action Plans*, 120. WorldFish Center Conference Proceedings 73, Penang: WorldFish Center.

Ponzoni, R., N.H. Nguyen, H.L. Khaw, N. Kamaruzzaman, A. Hamzah, K.R. Abu Bakar, and H.Y. Yee. 2008. *Genetic improvement of Nile tilapia (Oreochromis niloticus) – present and future*, 33–52. In Proceedings of the Eight International Symposium on Tilapia in Aquaculture, eds. H. Elghobashy, K. Fitzsimmons, A.S. Diab. Egypt, 12–14 October 2008, vol. 1.

Pullin, R.S.V. 1983. Choice of tilapia species for aquaculture. In *International symposium on Tilapia in Aquaculture*, eds. L. Sishelson, and Z.Yaron, 4–76. Nazareth, Israel, 8–13 May 1983. Tel Aviv: Tel Aviv University.

Pullin, R.S.V., and J.B. Capili. 1988. Genetic improvement of tilapias: Problems and prospects. In *The Second International Symposium on Tilapia in Aquaculture*, eds. R.S.V. Pullin, T. Bhukaswan, K. Tonguthai and J.L. Maclean, 259–266. ICLARM Conference Proceeding 15. Department of Fisheries, Bangkok, Thailand/International Center for Living Aquatic Resources Management, Manila, Philippines.

Pullin, R. 1994. Exotic species and genetically modified organisms in aquaculture and enhanced fisheries: ICLARM's position. *NAGA* 17: 20–24.

Pullin, R., M.L. Palomares, C. Casal, M. Dey, and D. Pauly.1997. Environmental impacts of tilapias. In *Tilapia Aquaculture*, ed. Kevin Fitzsimmons. Proceedings from the fourth international symposium on tilapia in aquaculture. vol. 2, 554–570. Ithaca, New York: Northeast Regional Agricultural Engineering Service Cooperative Extension.

Romana-Eguia, M.R., M. Ikeda, Z. Basiao, and N. Taniguchi. 2004. Genetic diversity in farmed Asian Nile and Red Tilapia stocks evaluated from microsatellite and mitochrondrial DNA analysis. *Aquaculture* 236: 131–150.

Rutten, M., H. Bovenhuis, and H. Komen. 2002. *Modelling fillet weight in Nile tilapia*. Abstract No. 06–13 of paper presented at the World Congress on Genetics applied to Livestock Production, 19–23 August, Montpelier, France.

Rutten, M., H. Bovenhuis, and H. Komen. 2004. Modelling fillet traits based on body measurements in three Nile tilapia strains *Oreochromis niloticus* L. *Aquaculture* 231: 113–122.

Silvestre, G.T., Garces, L.R., Stobutzki, I.C. Ahmed, M., Valmonte-Santos, R.A., Luna, C.Z. Lachika-Aliño, L., Munro, P. Christensen, V. and Pauly, D. (eds.) 2003. *Assessment management and future directions for coastal fisheries in Asian countries*, 110. WorldFish Center Conference Proceedings 67. Penang: WorldFish Center.

Taniguchi, N., J.M. Macaranas, and R.S.V. Pullin. 1985. Introgressive hybridization in cultured tilapia stocks in the Philippines. *Bulletin of the Japanese Society of Scientific Fisheries* 51: 1219–1224.

WorldFish Center. 2002. Report of INGA expert consultation on development of strategies and plans for dissemination of improved fish breeds to small-scale farmers. 4–7 June 2002, Pathumthani, Thailand: NAGRI

WorldFish Center. 2004. *GIFT* technology manual: An aid to tilapia selective breeding, WorldFish Center: Penang 46.

WorldFish Center. 2005. *GIFT Inventory*, 2. Penang: WorldFish Center.

WorldFish Center. 2006. Annual report 2005. Available at: ttp://www.worldfishcenter.org/e-AR2005/img/breeding.pdf

WorldFish Center. 2007. Policy on the transfer of genetically improved farmed tilapia (GIFT) for Asia to Africa by the WorldFish Center, 3. Penang: WorldFish Center.

WorldFish Center and PRIMEX, Inc. Philippines. 2007. Strategy for Sustainable Aquaculture Development for Poverty Reduction Project, Philippines (ADTA 4708-PHI). Final Report submitted to the Asian Development Bank (ABD). Penang: WorldFish Center.

Chapter 9
The Role of Exotics in Chinese Inland Aquaculture

Jiashou Liu and Zhongjie Li

Abstract China is not only the cradle of aquaculture, dating back to 2500 years ago, but in the context of modern aquaculture, it also leads global production. In China, inland aquaculture is the dominant component. However, with major socioeconomic changes that took place in the country since the beginning of the early 1980s, aquaculture has followed suite to meet consumer aspirations through a moderate shift into the culture of exotic species. The production of exotics increased from 780,000 tons in 1998 to 2.5 million tons in 2006, accounting for 5.9 and 11.7% of the total, respectively. Some exotic species have come to play an important role both in aquaculture production and economy, such as the channel catfish, tilapia, red swamp crayfish, sturgeons, and Pacific white shrimp. In a nutshell, the aquaculture production of exotics exceeds that of all continents other than Asia, emphasising the role of exotics in aquaculture in China.

The role of exotics in aquaculture is a much debated issue, if not controversial, particularly of their impacts from a biodiversity viewpoint. The exotics species in aquaculture in China is not known to have brought about such impacts to date. In view of the wide use of exotics in global aquaculture, it is most relevant to bring to light the success of exotics in Chinese inland aquaculture, the global leader.

9.1 Introduction

Food fish represent an important source of animal protein for many developing countries. More than one-third of animal protein in the diet of the Chinese people is sourced from fish. Fish is particularly important for the diet and nutrition of the poor, as it is one of the affordable forms of animal protein. Aquaculture has become an important industry for food security through the provision of income and employment in China, which is the global leader in aquaculture production. In 2006,

J. Liu (✉) and Z. Li
Institute of Hydrobiology, Chinese Academy of Sciences, 7 South Donghu Road, Wuhan 430072, China
e-mail: jsliu@ihb.ac.cn

S.S. De Silva and F.B. Davy (eds.), *Success Stories in Asian Aquaculture*,

the Chinese aquatic products reached 52.9 million tons, of which 36.9 million tons (i.e., 70%) were from aquaculture (FBMA 2007), representing over 70% of the global aquaculture production. It is estimated that 12.59 million people are engaged in the fisheries and aquaculture industry, of which 56% are in full-time employment (FBMA 2007). The average income generated through employment in the industry exceeds that of agriculture, with fisheries and aquaculture workers earning an average annual income of 5,816 RMB yuan in 2005 (6.8 RMB = 1 US$).

China has vast freshwater resources covering 18.2 million ha (1.9% of inland territory). Aquaculture industry plays a fundamental role in guaranteeing the economic and social development of China, improving peoples' living standards and maintaining social stability. Fisheries production accounts for 11% of the total output value of the agricultural industry, and contributed 3% to the total GDP in 2006. In China, 80% of fish production comes from freshwater aquaculture (FBMA 2007). The major cultured species are carps, especially the four major carps, i.e., silver carp (*Hypophthalmichthys molitrix*), grass carp (*Ctenopharyngodon idella*), bighead carp (*Aristichys nobilis*), and black carp (*Mylopharyngodon piceus*), and some other higher valued species, such as the mandarin fish *Siniperca chuatsi*, snakehead *Channa argus*, sturgeons, tilapia, and catfishes.

Diversification of cultured species is one of the goals of the aquaculture development in China. This, coupled with the nature of being hard working and the desire to indulge in food types of Chinese people, has resulted in the introduction of 129 aquatic species into China (Wang and Cao 2006), including finfish species such as tilapias, sturgeons, catfishes, trout, and salmon, and shellfishes such as crayfish and prawns. The production of exotics increased from 780,000 tons in 1998 to 2.5 million tons in 2006, representing 5.9 and 11.7% of total inland aquaculture production in China, respectively (Fig. 9.1). Some exotic species have come to play an important role both in aquaculture production and in the economy.

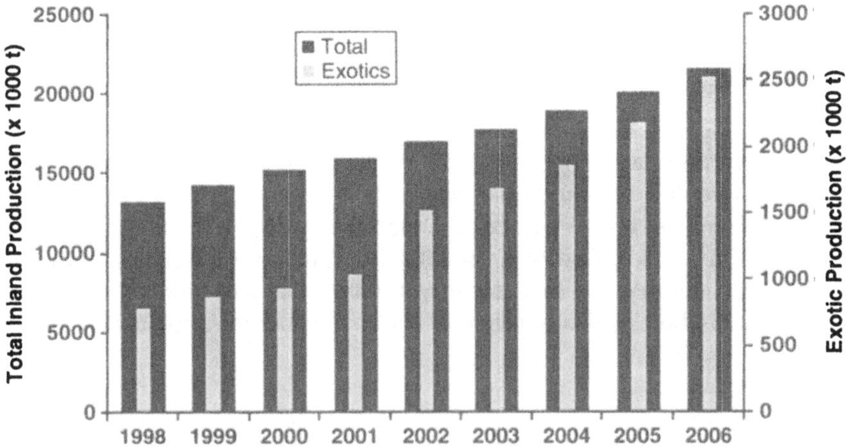

Fig. 9.1 Total inland and exotic aquaculture production in China

In general, it has been suggested by some that exotic species in aquaculture has been one of the primary causes for loss of aquatic biodiversity (see De Silva et al. 2009 for a review). However, it has been pointed out that explicit evidence for such impacts is not readily available, when dealt with on a case by case basis (e.g., De Silva et al. 2004). The role of exotic species in aquaculture has been dealt on global (Gozlan 2008) and regional scales (De Silva et al. 2006; Turchini and De Silva 2008) relatively recently. All of the above studies tended to emphasise on the importance of exotic species in aquaculture, and the overall consensus arrived was that these species have made a major contribution to food fish production and contributed to food security at large. In the present instance, China is taken as a case study where exotic species are becoming increasingly important in aquaculture. Accordingly, the role of some successful exotic species in aquaculture development in China is described in this Chapter, and lessons and suggestions from these experiences are discussed.

9.2 Aquaculture of Exotic Species in China

Although there have been many introductions of exotic species for aquaculture purposes, not all introductions are successful, as expected. In this document, only a few examples are brought forward to indicate the success and the reasons behind such successes.

9.2.1 Exotic Finfish in Aquaculture

Of the many finfish species introduced for aquaculture purposes in China, three species/species groups stand out, from the viewpoint of production as well as general and socioeconomic impacts. The production of all three groups and the percent contribution of each to overall global production are shown in Figs 9.2 and 9.3, respectively. The highest production is that of Nile tilapia (*Oreochromis niloticus*), a warm water species group of African origin, and it accounted for 65% of global production of the group, followed by channel catfish (*Ictalurus punctatus*), a species of North American origin, and by six species of sturgeon (Russian sturgeon *Acipenser gueldenstaedti*, Siberian sturgeon *Acipenser baeri*, *Acipenser ruthenus*, *Huso huso*, the American paddlefish *Polyodon spathula*, and *A. gueldenstaedti*, *A. ruthenus* and hybrid sturgeon the latter three being introduced into Dalian, China in 1993).

In respect of all three species groups, the trend appears to be one of increasing production. Moreover, China has gained dominance in the proportion of the global production of these groups, and in the case of sturgeon (Chen et al. 2008), it is nearly 85% of global aquaculture production. With regard to latter, however, the status is a little more complex as some countries culture sturgeon for its caviar (Fig. 9.3). The increasing importance of such species as plausible aquaculture species could indirectly have an influence on global aquaculture production of these groups *per se*,

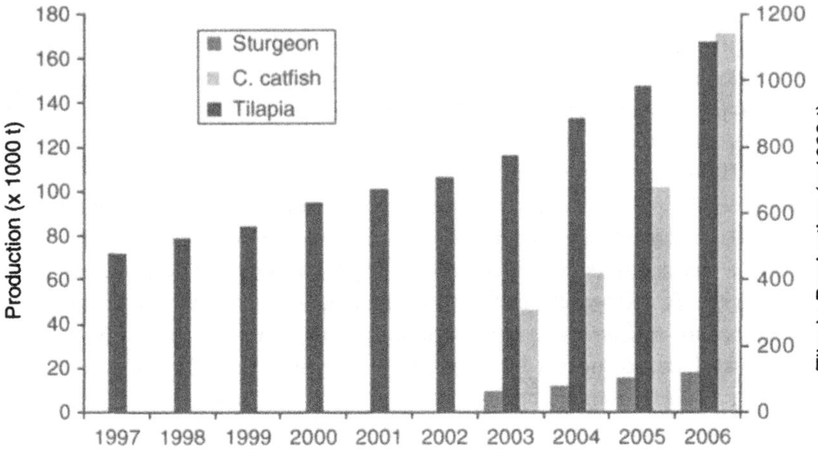

Fig. 9.2 Trends in production of three exotic finfish species in China

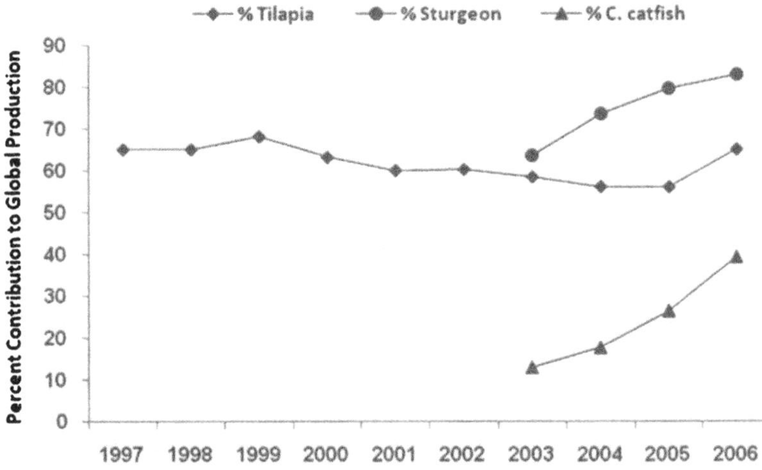

Fig. 9.3 The percent contribution to global aquaculture production of three exotic species in China

through the provision of a new commodity in the market place, thereby inducing other countries, particularly in those where such species occur naturally.

The spatial distribution of inland aquaculture based on exotic species is very distinct. This aspect is best exemplified by considering the geographic distribution of aquaculture of such species/species groups, based on 2006 production of red swamp crayfish and channel catfish (Fig. 9.4). It is evident that most of the aquaculture practices of exotic species in China occur in southern and central China, and are based on tropical and sub-tropical species, certain provinces leading such production.

Tilapias is one of the earliest exotic species groups introduced into China. More than eight species of tilapia were introduced into China (Li et al. 2007), and

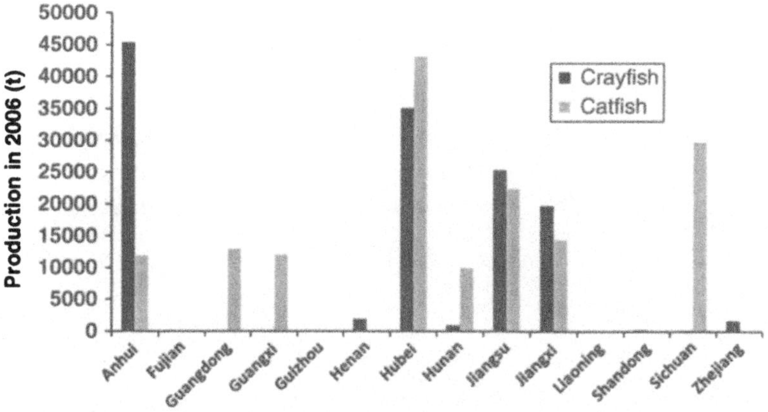

Fig. 9.4 The production of exotic crayfish and channel catfish in the different provinces in 2006

O. niloticus (Nile tilapia) is the most commonly cultured. From 1997 to 2006, tilapia production increased from 480,000 tons to 1.11 million tons in China (Fig. 9.2), with a yearly increment rate of over 10%. It ranks the sixth in inland aquaculture production in China after the indigenous species, such as silver carp, bighead carp, grass carp, common carp (*Cyprinus carpio*), and crucian carp (*Carassius auratus*) (FBMA 2007). China is the largest producer of tilapia in the world, primarily in the south-eastern coast, including Guangdong, Guangxi, Hainan, and Fujian provinces (Yang 2008).

China also produces most of tilapia fingerlings in the world, for example, it produced 14.4 and 15.4 billion tilapia fingerlings in 2005 and 2006, respectively (FBMA 2007). China is also the largest consumer as well as the exporter of tilapia in the world. In 2006, 375,000 tons of tilapia were exported, presenting 33.8% of the total production valued at US$ 0.368 billion, and 2/3 of tilapia was consumed in the country (Yang 2008).

On the other hand, sturgeon culture in China is more recent. Development of large scale artificial propagation techniques of sturgeons and a ready local market for the produce is thought to have triggered sturgeon culture in China. Now sturgeon aquaculture is found in more than 15 provinces, such as Guangdong, Hubei, Fujian, Jiangsu, Zhejiang, Jiangsu, and Beijing. In 2000, sturgeon production reached 4,000 tons in China and increased to 17,424 tons in 2006 (Fig. 9.2), representing more than 25% of the global production and currently accounting for more than 80% of the global production (Fig. 9.3). What is noteworthy is the rapidity in which all these exotic species began to make an impact not only on Chinese aquaculture production, but also on the share in global production.

One of the most recent introductions is the American paddlefish (Plate 9.1), the only species of the family Polydondidae that is cultured (Liu and Yu 1990). It was introduced into China in 1988 (He 1999). After more than 10 years of scientific study and culture, artificial propagation of the fish succeeded for the first time in

Plate 9.1 Broodstock of American paddlefish used for artificial propagation

Yichang, China (Tian and Wang 2001). Now the paddlefish is cultured in reservoirs, lakes, and ponds in more than 23 provinces of China.

9.2.2 Culture of Exotic Crustaceans

The two main exotic crustacean species cultured in China are the red swamp crayfish *Procambarus clarki* native to North America and introduced into China from Japan in 1929 (Wu and Gao 2008), and the Pacific white shrimp *Penaeus vannamei* introduced from the Marine Science Institute of Texas University in 1988. The culture of the latter is thought to have opened a new chapter in aquaculture in China when by 1994, hatchery techniques for the shrimp were established and commercial culture of the species began in the late 1990s.

Though the red swamp crayfish was introduced into China much earlier than other aquatic species, development of the culture industry has taken place only in the recent years driven by innovation in the cooking methods and establishment of export markets. In 2006, the production reached 130,000 tons in mainland China (FBMA 2007), and the main production occurs in Anhui, Hubei, Jiangsu, and Jiangxi provinces (Fig. 9.4). Crayfish are widely cultured in many water bodies, including ponds, rice fields, and lakes. In rice fields, the production could be 2,250–3,000 kg ha^{-1}. In ponds, the production could be double that in rice fields.

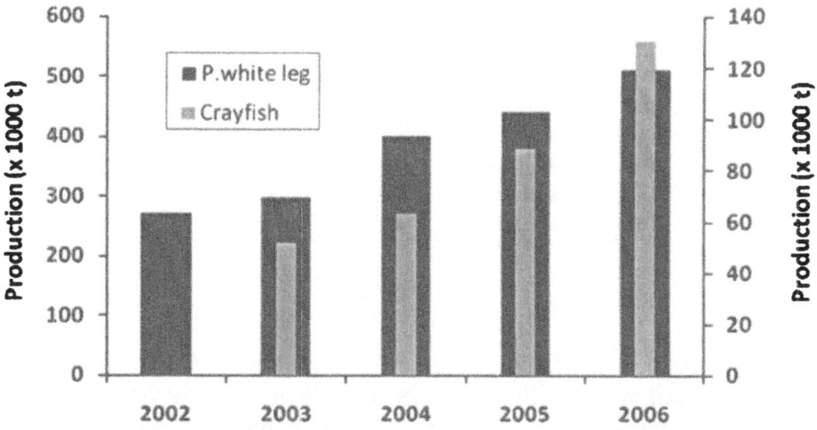

Fig. 9.5 Production of the Pacific white shrimp and red swamp crayfish in China

In 2006, the aquaculture area for the crayfish reached nearly 40,000 ha with a production of 35,000 tons in Hubei province, presenting 30% of the country's total production (Fig. 9.5).

In the case of *P. vannamei,* a series of experiments were conducted in a range of salinities (Pong et al. 2005), and based on such findings, gradually its culture was extended inland. The pond production in seawater could reach 7500 kg ha^{-1} (Shen et al. 2004). Freshwater culture of *P. vannamei* in China reached approximately 510,070 tons in 2006 (Table 9.1), accounting for more than half of China's total inland shrimp/prawn production (Table 9.1).

In addition to the inland production, in marine waters, *P. vannamei* production reached 512,019 tons in 2006, representing 69.9% of total marine shrimp production in mainland China. Postlarvae production of *P. vannamei* reached 25.11 billion in 2005 and 49.24 billion in 2006, which ensured the high production of the shrimp both in freshwater and marine water.

Overall, the contribution of *P. vannamei* production from China to global production is on the decline (Fig. 9.3). This, however, is not a result of a reduction in production *per se*, but due to an increased production in other Asian countries, such as in Thailand and Vietnam.

9.3 Markets and Processing Industry Developments

The development of the culture of exotic finfish and crustaceans has also resulted in the development of export markets and associated processing sectors. For example, from 2003 onwards, the number of processing plants increased fast and the export of channel catfish fillet to the United States increased from 326 tons in 2003 to

Table 9.1 Comparison of *P. vannamei* and total inland shrimp production (tons/year) in China

Year	*P. vannamei*	Total shrimp	Percentage
2002	272,980	415,000	65.8
2003	296,312	690,218	42.9
2004	400,714	803,679	49.9
2005	440,791	847,256	52.0
2006	510,070	981,755	52.0

6,821 tons in 2006. It was estimated the fillet export to the United States reached 28,800 tons valued at 114 million US $ in 2007 (Wang and Liu 2007).

On the other hand, China has been a leading exporter of processed cultured tilapias for over a decade. USA is the leading importer of tilapia, which is the freshwater fish commodity that is imported in highest quantity to the country. Tilapia is imported in three forms to the USA: frozen whole, frozen, and fresh fillets. China has the highest market share in the first two commodities, and it has been on the increase and constitutes a very well-established market share (Fig. 9.6), in spite of the high growth in tilapia culture in some South American nations, which tend to have the major share of the fresh fillet imports to the USA.

Jiangsu province was the biggest producer and exporter of the red swamp crayfish before 2003. There were more than 60 processing factories for crayfish in the province, producing 90% of the total crayfish exports. However, in recent years, Hubei province became the biggest exporter for crayfish in China. In Hubei, there are 15 processing factories for crayfish with a capacity of 200,000 tons per year. In 2006, 12,000 tons of processed crayfish were exported from the province, amounting to US $ 82 million.

In essence, China's export of fishery products is primarily based on a handful of cultured exotic species, including the shrimp, *P. vannamei*. These export industries provide very significant numbers of employment opportunities, mainly to rural women, as in the case of the rest of the region.

9.4 Possible Reasons for the Success of Aquaculture of Exotic Species

The exotic species that have been successful in aquaculture in China are essentially tropical and/or subtropical species, with relatively high growth rates and other attributes of the lifecycles that are favorable for aquaculture, and also of desirable flesh quality relished by the local consumers, and/or have a ready export market such as for tilapias and shrimp.

Although there was no rigorous risk assessment process involved before the introductions, particularly the early introductions, investment on research and technology

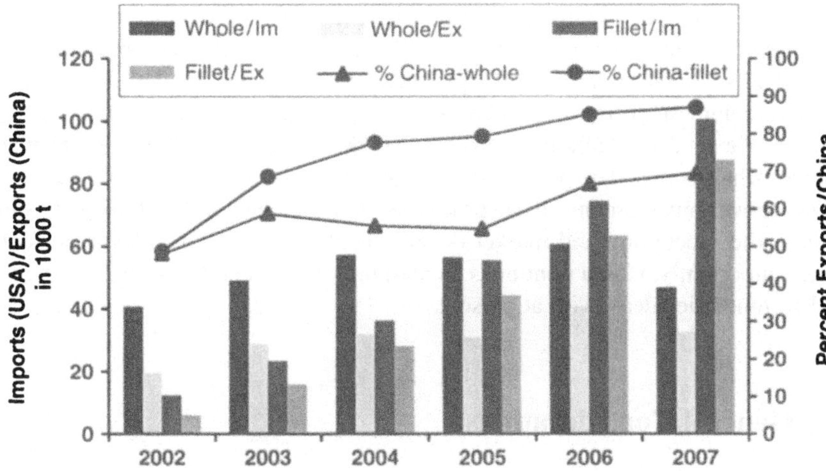

Fig. 9.6 The export quantities of frozen whole and filleted tilapia into USA, the amount of imports of each to USA from China, and the percent imports from China for each of the commodities. Based on data from: http://www.eurofish.dk/indexSub.php?id=3541&easysitestatid=-61206804; Helga Josupeit © 2008 FAO

development was one of the priorities for aquaculture of exotic species. For example, in the case of channel catfish that was first introduced from Japan in 1978 (Wang and Liu 2007), work was undertaken on reproductive aspects and fry rearing (Huang 2008a, b), on the developments of suitable diets based on basic nutritional research (Li 2002; Cheng et al. 2005), most effective feeding regimes (Huang 2008c), and common diseases and treatments, including Chinese herbal treatments (He 2006; Tao et al. 2008).

The above was followed by developing suitable culture techniques and management regimes for each of the species. For example, in 2003, Hubei province published the Criteria for Cage Culture of Channel Catfish (He et al. 2003), followed by publication of the Criteria for Channel Catfish Aquaculture by the Chinese Ministry of Agriculture (CMA 2004). These helped to standardize aquaculture practices of channel catfish, decrease the occurrence of diseases, and improve food safety. Accordingly, cage culture of channel catfish is more predominant, and is mainly conducted in reservoirs and cages (size range from 4×4×2.5 m to 5×5×2.5 m; Luo et al. 2008), and has brought about large livelihood opportunities and improved the living standards of displaced people (Hu and Peng 2008).

As for tilapia, because it can accept a wide range of food types, it was adopted for pond culture by even remote rural farmers such as in Chitan County, Fujian Province. Two basic systems, monoculture and polyculture, are practiced for production of table sized tilapia in ponds by rural fish farmers. Both culture systems produce high yields and good economic returns. In the polyculture system, tilapia is stocked as the main species at a density of 30,000–37,500 individuals/ha together with silver carp (3,750 individual/ha), bighead carp (450–600 individuals/ha), grass carp (750 individuals/ha), and common carp (150 individuals/ha). The production may reach 10,000–13,000 kg ha^{-1} with tilapia accounting for 70–80%. The other practice is to culture domestic carps as the main species and tilapia as the auxiliary species. In this case, the total production may reach 13,000 kg ha^{-1} with 7,000–9,000 tilapia fingerlings/ha stocked. The production of tilapia may reach over 2,250 kg ha^{-1}.

The above developments have proceeded almost hand in hand with developments in the processing and market developments, aspects which were presented earlier. Importantly, China continues to maintain its market dominance of the cultured commodities dealt with at present.

9.5 General Considerations

The exotic species that are successfully dealt with presently are all tropical and/or subtropical species, with considerable market potential. The successes witnessed have resulted from suitable adaptation of each to culture conditions in China, backed up by effective research and dissemination of findings. In all instances, the authorities have taken appropriate action to popularize the species with culture and market potential, local and/or foreign

From the experiences of successful production of some exotic species in China, we learnt that studies related to local practices are very important. Biology and aquaculture techniques of the introduced species need to be studied thoroughly under local conditions even though some information of the species is available in places of origin. For example, in respect of channel catfish culture, the facilities used in the United States were more sophisticated and not directly adaptable to Chinese rural conditions. Simple facilities created in China for spawning and hatching are also equally efficient, widely practiced, and easy to adopt, resulting in large scale production

The use of local ingredients in diets also helped to reduce feed costs. This is the key to good economic results for farmers. A wide range of feed ingredients is available in China, and as such, producing suitable diet for imported species is not necessarily a constraint, provided it is backed up by research. In remote rural areas, farmers could use local ingredients to feed their fish and produce high value fish like in the case of aquaculture of channel catfish, crayfish, tilapia, and prawns, and improve the economic returns.

The success of an exotic species in production must have a wide market accessibility. When channel catfish was first introduced into Geheyan Reservoir, Changyang County, Hubei Province for cage culture in the 1990s, the market was very limited. The officers in the county invested heavily on personnel resources to bring about public awareness of the new commodity. Now the fish is called "Qingjiang Fish," which has become a famous brand in Hubei Province and beyond. Geheyan Reservoir has also become the biggest base for channel catfish production. Now the internal market of channel catfish is also developed.

China is heavily populated. Local consumption of aquaculture produce tantamount to a huge market. The red swamp crayfish was one of the earliest exotic species in China. But aquaculture of the crayfish did not receive attention, and in fact, was even thought as a pest species, since there is perception that it might burrow into dams. However, when the dish "chilly crayfish" came to the table, most Chinese accepted the dish, which brought about a large scaled production of crayfish.

9.6 Risk Assessment and Control

Though exotic species in aquaculture in China bring about vast economic returns and provide huge employment opportunities, its risk to local fish diversity and ecosystem is a major concern (Ding 2007; Hu et al. 2006; Wang and Cao 2006; Bu et al. 2008; Wang et al. 2002).

Admittedly, China has insufficient knowledge on systematic studies on exotic species. So far, there is no complete exotic species list, though 16 species were recorded as invasive by the government. None of the 129 aquatic exotic species reported (Wang and Cao 2006) was included in these 16 invasive listed.

Thus far, China has no specific laws or regulations for the management of exotic species, and no strict risk assessment mechanisms, comprehensive control mechanisms, and trace mechanism for exotic species.

However, there are a lot of laws and regulations related to exotic species control, such as the Environmental Protection Law of the People's Republic of China, the Marine Environment Protection Law of the People's Republic of China, the Agriculture Law of the People's Republic of China, Fisheries Law of the Peoples Republic of China, and the Law of the People's Republic of China on the Entry and Exit Animal and Plant Quarantine. For instance, the Environmental Protection Law is related to the protection of environmental organisms, but not to exotic species invasion and relevant control. In the Agriculture Law of the People's Republic of China, introduction of exotic species should be examined, approved, and registered, and relevant safety measures need to be taken.

9.7 Administration Institutions for Exotic Species

Many institutions are involved with the administration of exotic species, including General Administration of Quality Supervision, Inspection and Quarantine of China, the Ministry of Environment Protection, Ministry of Agriculture, and State Forestry Bureau, but there is no united administration institution for exotic species. In the current context, it may be desirable to have one single authority dealing with exotic species and their impacts, and regulation of new introductions.

In spite of the relative lack of stringent procedures in regard to species introduction into China, in a manner China has been lucky in that none of the aquatic exotics species introduced for aquaculture purpose has been invasive, and has not impacted adversely on biodiversity. Perhaps, this is mainly due to these species not been able to establish natural reproductive populations because of the extreme climates experienced in the regions in China into which these species have been introduced into. However, this is no reason for complacency. The fact that China has recognized all the complexities of introductions and the need to set up the required administrative and institutional backing is most welcome.

References

Bu, Y., C. Xie, X. Liu, and Y. Ren. 2008. Environmental risk assessment of introduced species. *Reservoir Fisheries* 28: 82–84. In Chinese.

Chen, P., B. Xiong, G. Lu, J. Wang, and Z. Dai. 2008. Aquaculture status and prospects of the American paddlefish. *Acta Agriculture Jiangxi* 20: 88–89. In Chinese.

CMA (Chinese Ministry of Agriculture). 2004. *The criterion of channel catfish aquaculture.* China Agricultural Press (in Chinese): Beijing.

De Silva, S.S., R.P. Subasinghe, D.M. Bartley, and A. Lowther. 2004. Tilapias as alien aquatics in Asia and the Pacific: A review. *FAO Fisheries Technical Paper* 453: 65 pp.

De Silva, S.S., T.T.T. Nguyen, N.W. Abery, and U.S. Amarasinghe. 2006. An evaluation of the role and impacts of alien finfish in Asian inland aquaculture. *Aquaculture Research* 37: 1–17.

De Silva, S.S., Nguyen, Thuy T.T., Turchini G.M., Amarasinghe U.S., Abery N.W., 2009. Alien species in aquaculture and biodiversity: a paradox in food production. *Ambio* 38: 24–28.

Ding, D. 2007. Healthy culture of big sized channel catfish fingerlings in cages. *Inland Fisheries* 2: 22–23. In Chinese.

FBMA (Fisheries Bureau of Ministry of Agriculture). 2007. *China fishery statistical yearbook.* China Agriculture Press: Beijing (in Chinese).

Gozlan, R.E. 2008. Introduction of non-native freshwater finfish: Is it all bad? *Fish and Fisheries* 9: 106–115.

He, Y. 1999. Development prospects of the American paddlefish and aquaculture techniques. *China Fisheries* 4: 25–27. In Chinese.

He, G. 2006. Control of gill erosion disease of channel catfish induced by the syndrome of liver and gall. *Reservoir Fisheries* 4: 92–93. In Chinese.

He, G., S. Hu, C. Li, L. Zou, and G. Zhao. 2003. *The criterion for cage culture of channel catfish.* Hubei Provincial Bureau of Quality and Technique Surveillance (in Chinese): Wuhan.

Hu, S. and J. Peng. 2008. Status and development trend of channel catfish industry in reservoirs of the Qingjiang River, Hubei. *Reservoir Fisheries* 28: 108–109. In Chinese.

Huang, A. 2008a. Artificial propagation of the channel catfish and environment-friendly rearing of the fry and fingerlings. *Scientific Fish Farming* 4: 14–16. In Chinese.

Huang, A. 2008b. Artificial propagation of the channel catfish and environment-friendly rearing of the fry and fingerlings. *Scientific Fish Farming* 5: 14–16. In Chinese.

Huang, A. 2008c. Artificial propagation of the channel catfish and environment-friendly rearing of the fry and fingerlings. *Scientific Fish Farming* 6: 14–16. In Chinese.

Li, A. 2002. *Nutrition and feeds of aquatic organisms.* China Agricultural Press: Beijing (in Chinese).

Li, X., A. Li, X. Peng, and G. Cheng. 2007. Introduction and stock resources protection of tilapia in China. *Henan Fisheries* 1: 3–6. In Chinese.

Liu, J. and Z. Yu. 1990. The American paddlefish and its fisheries in the U. S. A. *Acta Hydrobiologia* 14(1): 75–83. In Chinese.

Luo, H., H. Lu, and H. Wang. 2008. Healthy production of channel catfish in reservoir cages. *Inland Fisheries* 4: 45–46. In Chinese.

Pong, J., F. Zhang, G. Xiao, and Z. Cui. 2005. A study on breeding technique of *Penaeus vannamei* and disease occurrence of adult shrimp culture. *Modern Fisheries Information* 4: 12–14. In Chinese.

Shen, D., L. Si, and J. Wang. 2004. Test report on intensive culture of *Penaeus vannamei* in ponds. *Modern Fisheries Information* 6: 12–14. In Chinese.

Tao, R., D. Peng, Y. Zhang, A. An, and Z. Liu. 2008. Studies on the bacteria in for haemorrhage disease of the channel catfish. *Inland Fisheries* 3: 17–18. In Chinese.

Tian, Z. and Z. Wang. 2001. The artificial propagation of the American paddlefish was successful in Yichang, China. *Reservoir Fisheries* 21(3): 3. In Chinese.

Turchini G.M., De Silva, S.S., 2008. Bio-economical and ethical impacts of alien finfish culture in European inland waters. *Aquaculture International*, 16: 243–272.

Wang, Y. and W. Cao. 2006. The strategies of aquatic exotic species in China. *Journal of Agro-Environment Science* 1: 14–19. In Chinese.

Wang, Y. and F. Liu. 2007. Channel catfish industry development status and strategies in China. *China Fisheries* 12: 14–16. In Chinese.

Wang, Z., H. Liu, and Z. Lu. 2002. Aquaculture techniques of the American paddlefish using formulated diets. *Beijing Fisheries* 6: 17. In Chinese.

Wu, T. and P. Gao. 2008. Status and development prospects of the freshwater red swamp crayfish. *Inland Fisheries* 2: 15–17. In Chinese.

Yang, B. 2008. The increment growth of tilapia production will decrease in 2008. *Inland Fisheries* 2: 11. In Chinese.

Chapter 10
Synthesis and Lessons Learned

Sena S. De Silva, F. Brian Davy, and Michael J. Phillips

Abstract This synthesis now returns our discussions to some of the ideas and questions that were posed in the introduction by reexamining them from the point of view of the lessons learned from the selected cases. These issues are then further examined as part of a broader set of lessons learned around successes in aquaculture with the aim of developing improved guidance or influencing strategies around possible steps to follow; steps in the further development of this set of ideas, and indeed in the sector as a whole.

The original purposes in developing the case studies presented before were to:

- highlight experiences in aquaculture development which have led to positive social change and negligible ecosystem impacts
- provide a better understanding of the factors and approaches leading to sustainable aquaculture growth combined with positive societal changes
- show that small scale farmers have incentives for and can act responsibly
- provide a direction for future policy change and collective improvement to ensure sustainability of the sector, and
- attempt to mitigate negative public perceptions of aquaculture

We feel that the cases do add flesh to our purposes bearing in mind that these are part of a longer term process involving issues which need continued attention.

10.1 An Evolving View of Success

At the outset, it is important to attempt to determine or define what success is and when it is apparent that success could be seen or measured and evaluated in its many forms. For the present purposes, it was suggested that success can be seen as

S.S. De Silva(✉)
Network of Aquaculture Centres in Asia-Pacific, PO Box 1040 Kasetsart Campus, Bangkok, 10903, Thailand
e-mail: sena.desilva@enaca.org

S.S. De Silva and F.B. Davy (eds.), *Success Stories in Asian Aquaculture*,

part of an evolving view of sustainability, reflecting the thinking that current uses of a system should not reduce and/or impair its ability to continue to provide services to the future generations. Sustainability essentially rests on three pillars, viz., ecological, sociological, and economic, but on many an occasion, impacted upon by geo-political factors.

In striving for sustainability, there has been a gradual convergence of aquaculture into fisheries and fisheries into a wider set of ecosystem-based thinking. Fisheries thinking, however, as one looks at it, continues to evolve, having primarily moved toward ecological considerations such as biodiversity and ecosystem health during the 80s and 90s, with more recent moves into participation and related social issues in the last decade or more. This can be seen as a drive toward a more multidisciplinary thinking and a holistic approach. Aquaculture, on the other hand, traditionally has been based on a strong sense of private ownership (local/national) and economic resilience, building on much older animal husbandry/farming systems linkages. In the current day and age, both fisheries and aquaculture increasingly have to come to terms with ecological sustainability closely linked to economic viability and social responsibility.

10.2 A Vision for Sustainable Aquaculture

Globally, aquatic foods, which are of high nutritional value, contribute 20% or more to the total animal protein intake, and in developing countries, its contribution is even higher. Overall, it is estimated that more than 42 million people work directly in the fisheries sector, with the great majority in developing countries. Adding those who work in associated processing, marketing, distribution, and supply industries, the sector supports several hundred million livelihoods.

Over the last 30 years, there have been significant and notable changes in the fishery sector; foremost amongt these are the shifts, in consumption and production patterns, from a developed country dominance to a developing country dominance. Aquaculture, though a tradition for many a millennia, has come to contribute significantly to the human food basket only in the last two to three decades. It currently accounts for nearly 50% of all global food fish consumption, and is expected to contribute upto 60–65% by the year 2020. This significant impact on the human food basket has occurred in a scenario where the global population has been increasing, as well as becoming increasingly urbanized. On the other hand, in this day and age, public perceptions and awareness on ecosystem health are taking center stage and many forms of new development are being criticized and dismissed, unless these are proven to be sustainable and impact the environment minimally. Sustainable aquaculture needs to be well-positioned amid these changing contextual dynamics.

Apart from the direct contribution of aquaculture to the human food basket, it has equally contributed to food security and poverty alleviation through the provision of many millions of livelihoods dependent directly and indirectly on activities of the sector. In many leading aquaculture nations, it has begun to bypass the contribution to the GDP from capture fisheries, and equally its processing sector developments are a significant avenue for employment of rural women, empowering their status in the households and in the community at large.

Aquaculture has now reached an important position as a globally significant contributor to food production. However, this development has had its critics, both from a primary resource use as well as consumer and social responsibility viewpoints. A closer scrutiny of these criticisms will reveal that these are mainly directed toward those aquafarming systems that tend to produce high valued, carnivorous species, such as marine finfish, mainly salmonids, and shrimp. It also raises the age old issue of why should the world endorse a food production system that converts what is perceived as low quality protein to high quality protein that is within the reach of only the rich, thereby depriving the millions of poor who are unable to obtain the basic animal protein needs. This ethical question in the use of nearly 25% of capture fisheries of low valued fish/trash fish/forage fish for reduction purposes primarily to produce ingredients for animal feeds including feeding fish (rather than a direct food source for the millions of hungry and poor) will remain in the forefront for a long time to come (Aldhous 2004; Powell 2003). Needless to say, this contention has been further exacerbated by the recent revelations that a significant proportion of the above resources are being channelled into nonhuman food production (De Silva and Turchini 2008). Also in terms of ecological efficiency questions, it is important to stress that global aquaculture production is dominated by the culture of organisms that feed lower in the food chain (e.g., bivalves, seaweeds, the great bulk of finfish species), and this trend is likely to continue well into the foreseeable future. Many of the cases cited earlier in this compendium follow this trend (e.g., filter feeding carp culture in cages in Nepal, China rice fish, CBF in Sri Lanka). Whether consumer preferences for such species will continue and indeed increase in the future remains to be seen.

Previous efforts in addressing some of the sustainability questions are briefly summarized in The Bangkok Declaration and Strategy of April 2000. Therein a declaration was made and a further 17 key strategies were proposed toward attaining sustainable development needs, and corresponding implementation mechanisms(s) were developed. This strategy has sought to address those issues related to Aquaculture Development Beyond (2000) in the new millennium; a process through which the above mentioned 17 key elements are now being developed and implemented to varying degrees with varying degrees of success. In summary, significant progress has been made on the sustainability issue, but more needs to be done. The above thinking has guided much of the present endeavor to document success stories, and we have sought to summarize this thinking in Box 10.1.

Box 10.1 A Vision for Sustainable Aquaculture/Fisheries (Developed from the Bangkok Declaration)

Healthy Ecosystems

- Decline of marine and freshwater fish stocks is reversed through reduced commercial pressures, habitat protection, and ecosystem restoration.
- Commercial aquatic products are produced largely through aquaculture systems that are ecologically more benign and affordable. This means cultivation that minimizes the use of chemicals, drugs, and hormones, minimizes impacts on ecosystems, and reduces use of animal protein in feed (is a net producer of aquatic protein).
- Pollution abatement in aquatic ecosystems, including from upstream sources.
- Ecosystem goods and services (EGS) are defined and accessed, and protected locally.

Sustainable Livelihoods and Sustainable Enterprises

- Benefits flow more equitably to users. Smallholder fishers and aquaculturists are treated with respect and dignity, and are assured in their livelihoods.
- Assured access to resources, reasonable terms for credit, and markets for small-scale producers.
- Businesses emerge, thrive, and invest within local communities.
- Risks are reduced through better forecasting and access to key information, and good communications.

Improved Governance

- Enabling frameworks in place for attaining sustainability with funding to cover the implementation of relevant sustainability strategies and actions, including monitoring and effective enforcement.
- Approaches are more participatory and multidisciplinary – where the knowledge is communicated and able to be used to improve sustainability and for the benefit of local producers and communities.
- Management costs and participation in decision making are shared between producers, industry, and local, national and regional/international governments and agencies.
- Appropriate growth and innovation strategies.

Organizational Change

Organizations and staff with increased capacities to develop appropriate tools can implement adaptive learning-based strategies guiding sustainability processes in a collaborative fashion.

10.4 Measures of Success

This compendium is thought to be the first attempt at documenting success stories in aquaculture, and hopefully it will stimulate others to follow suit, particularly from other regions of the world and other aquafarming systems. The exercise has not been straight forward nor was it easy, especially when one considers the great diversity and rapid changes in species cultured, size, shape, intensity, resource usage, socio-economic benefits, among others, in aquaculture.

One criterion suggested in reviewing this material is that one element of success is resilience over time or perhaps what might be considered as the persistence factor. For instance, most of the chosen cases have been in operation as aquafarming systems for a decade or more (and as explained in the rice-fish case in China that has been in operation for more than 2500 years) and all signs indicate that these will continue to be sustainable in the longer term and continue to contribute to the ever increasing human food fish needs, particularly from the viewpoint of meeting the animal protein needs. Resilience offers one important measure of success.

Also adaptive capacity or the ability of the production systems to evolve in a diversity of contexts and related issues of the "fit" of the production system in relation to such contexts is another key factor for success. This comparison is best exemplified by the two extreme cases illustrated by simple cage farming in Nepal and the rather explosive growth of catfish farming in Vietnam. In the former instance, a low productive and environmentally minimally perturbing system has been sustained for over two decades while continuing to provide improved livelihoods for the farmers than that they were engaged in prior to the impoundment of the Kulekhani reservoir. In Vietnam, on the other hand, the explosive growth of an aquafarming system, primarily built around the culture of "tra catfish," has succeeded in developing a pond-based production system tapping the traditional global demand for a "white fish" of modest market price. This farming activity, centered in the Mekong Delta, reached a production of 1.2 million tons of food fish in 2007, which generated a foreign exchange income of US$ 987 million and provides livelihood opportunities to more than 150,000 rural people, mostly women engaged in the associated processing sector. Both of these cases represent successes even though they reflect two extremes in absolute terms in the overall intensity of the practices, monetary value of the operations, as well as individual monetary gains.

Development of flexible supportive aquaculture services has been another very different example of success. The case of backyard hatcheries, primarily in Thailand, has enabled the Thai shrimp and mariculture sector more generally to maintain its global leadership for nearly a decade. This resilience is driven by mainly the small scale, low capital and operation costs of these hatcheries, which have permitted a switching behavior by farmers that has guided the continued development of this sector even through the turmoils of major viral disease outbreaks, market price fluctuations, and changing competitive market forces. This ability to make shifts to other appropriate species or even temporarily cease farming activity for one or more production cycles are the hallmarks of these systems. Farmers are the flexible backbone of Thai aquaculture – and it is not just hatcheries, but a host of interweaving factors that contribute to a chain of events.

Scientific and technological advances did place aquaculture on a firm footing three to four decades ago. For example, the sector became independent of wild caught seed stock in respect of many cultured species, and correspondingly, the management levels improved resulting intensification of the practices. It is also pertinent to point out that the science of aquaculture is relatively new and hence, lags far behind that of the terrestrial husbandry sector. Added to this is the fact that aquaculture deals with in excess of two hundred species/species groups, unlike the livestock sector, and makes it that much more difficult to make technological advances for all associated species/species groups; this would be most evident in the case of the level of genetic improvements that have been achieved in cultured aquatic organisms (see the GIFT tilapia case for more details on this issue). It was also apparent from the cases of success stories presented that technological advances alone were not adequate to attain and/or maintain sustainability, without associated changes in the grass roots organizations of farmers, particularly in view of the fact that the great bulk of aquafarming systems in Asia are small-scale: best defined as farmer owned, operated, and managed systems.

Understanding and guiding improved partnerships and related forms of social organization have been another key to success, for instance, in the revival and sustainability of small scale shrimp farming on the east coast of India, which is now being actively extended into many neighboring coastal states, as well as beyond India. This evolving and expanding cooperative-problem focussed approach assisted the small scale shrimp farmers to work together to deal with water quality, shrimp health, and now market access. This set of practices is becoming extremely sophisticated and complex, and has both empowered these cluster groups as well as promoted cost savings, improved market accessibility, and increased profitability.

Success in aquaculture in Asia continues to evolve. It is sometimes based on exotic and or alien species, even though such alien species have been a bone of contention by many an environmentalists, based on its perceived negative impacts on biodiversity. The dilemma on the use of alien species presents a challenging set of continuing questions around providing food vs. hunger. Needless to say that careful use of exotics, as in the case of China as well as elsewhere (De Silva et al. 2004), has contributed significantly in providing livelihood opportunities, food security, and poverty alleviation.

Overall, it is proposed to continue this initial process of identifying and building on success stories in aquaculture. Much remains to be done on identifying, reviewing, and challenging all stakeholders, which will help in furthering the development of positive lessons (as well as lessons from failures) of a wider variety of comparable aquaculture systems in the long run.

10.5 An Evolving "Success" Hypothesis

Early planning for the documentation of "success stories in aquaculture" started around a basic hypothesis that success in whatever form was not adequately known and/or understood, nor documented, especially in respect of this sector. It was also felt that the evolution of thinking in this sector has also changed given the increasing numbers of aquaculture stakeholders voicing concerns from an increasing set of organizational levels and viewpoints. Accordingly, the problems of scale and levels in the current exercise need to be better highlighted/ recognized, particularly in terms of the thinking around shared learning, improved dialogue, and related exchange processes. Therefore, continued success must be seen as part of a process of increasing dialogue well into the future.

Change is a given entity, but the frequency and speed of change are suggested to accelerate in the future. Therefore, one of the conclusions was that inadequate processes exist to develop or promote a shared understanding of aquaculture and its evolution to date, and more so related to managing its future growth. As suggested earlier, part of this problem relates to an inadequate understanding of the past evolution of aquaculture, and it is believed that some of the success stories cited in this compendium represent examples of sustainability and resilience that merit a more balanced and perhaps detailed examination by a wider audience. If sustainability changes are to continue, we need to make greater efforts to reach and involve this wider set of stakeholders. This compendium, some of its conclusions, and next steps are part of the efforts to provoke more discussion and exchange around this set of widening concerns.

10.6 Conditions Favoring Success

As mentioned earlier, demonstrated resilience or staying in power over a number of years, crises, and challenges has been a main measure of success. The successes discussed in this compendium have been developed, adapted, and persisted in one form or another over extended periods. For instance, the China rice-fish example that is suggested to have started some 2000 years ago as one of the most enduring cases overcoming constraints and socio-economic changes in many fronts. This lesson is in a way to say the obvious. However, in the process of achieving the successes, there are many issues that have been learned, some of them generic and others more specific to each situation. These lessons learned are worth some degree of consideration as these could inevitably be useful in future endeavors in aquaculture development.

The lessons learned could be broadly categorized under (at least) the following general headings:

- Enabling policy and legislative environment
- Empowerment of farmers and/or communities

- Effective linked institutions; partnering, communication, and dissemination systems
- Efficacy of resource usage
- Opportunistic behavioral changes

It was relatively surprising that in all instances, the successes were achieved with and may be even because of accompanying and relevant policy changes; changes that were a facilitator in the processes. Tracing the history of each case cited suggested that:

- often a trial and error process in which the changes that were needed to achieve success were not immediately apparent in the initial stages of the development process, and
- equally, the policy changes that facilitated the developments were often not under the purview of the authorities for aquaculture development *per se* (often representing serious cross organizational challenges). Macro-economic policy could have a major influence, which we tend not to be often aware of and/or take for granted – e.g., Vietnam catfish could not have grown so rapidly without "Doi Moi" policy (= open economic policy)

As development proceeded, aquaculturists had to tread on a fine line and convince other parties of the needs, and most importantly, show that the suggested changes would not impede and/or interfere with the processes that are already in place. In essence, the need for cross-fertilization and dialogue between sectors finally leads to a more holistic approach to resource use, empowerment of the users, and a move toward sustainable development. The institutional and policy environment that allows such communication and innovation – not to mention the underlying economic and cultural contexts – equally played an important role.

In Table 10.1, an attempt is made to sieve out the lessons learned across a cross-section of all the cases, and then to use these lessons to illustrate some of the evolving implications of these lessons learned and how these might be used to guide more effective change processes in the future (as part of the move toward increasing sustainability as shown in Table 10.1).

10.7 A Look into the Future

Following up from the ideas in Box 1, a number of issues for further analysis and follow-up action in relation to the developments in sustainable aquaculture were identified and are enumerated in the table (Table 10.1) and the following section.

- How change happens: including more detailed examination of the evolution of aquaculture within the rapidly changing global context around the demands of an expanding human population, which is becoming increasingly urbanized.
- The recent expansion of aquaculture and the fact that aquaculture is still a relatively new food production sector compared to others need to be considered in relation

Table 10.1 Conditions favouring success: generalized lessons learnt from the cases

Generalized lessons learnt	Example case study(ies)	Implications for future planning[a]
Resilient systems take time to create a successful change – often more than 10 years of trial and error learning in many cases	Demonstrated in most cases. e.g., CBF, GIFT	This lesson suggests that a long term commitment is required by most stakeholders, which has a variety of planning and particularly resource implications. Also this longer term trial and error testing seems to lead to more resilient production systems. *We see opportunities to use lessons learnt to speed up this process to cope with impending future changes*
Context and/or place is critically important	All cases	Problems and the need for change must be seen in their context. Stronger, more *participatory action and maximal engagement of all stakeholders need to start from the place-based problem identification stage directly involving the users*
Enabling policy, macro-economic conditions and legislative environment can be critical elements	For example, in CBF in Sri Lanka a new legal and policy environment was needed	Policy, however, can move and impact in a variety of directions. The policy shift leading to the banning in Sri Lanka of inland fisheries also obviously set this whole CBF process back greatly, *while its withdrawal and subsequent amendment of the Agrarian Act of 2000 facilitated the increased adoption of CBF and increased community confidence*
Empowerment of communities	Most cases, but the Nepal and Indian cases, demonstrate this issue very convincingly	Social-organizational changes are increasing in importance across all the cases suggesting that changes in organizations and stakeholders need to better understand this set of issues and appropriate capacities need to be available to guide these changes. *Aquaculture has been relatively slow to move into interdisciplinary thinking and a shift in this regard is vital to the sector's development and sustainability*
Effective institutional linkages and communication	CBF in Sri Lanka	Linkages across government departments (e.g., DAD and NAQDA in Sri Lanka), utilizing AEOs and across levels (local, national regional and global) are an increasing part of the networking and dialogue processes. *More effort needs to be expended to reach across levels of authority and associated stakeholder groups (cf. co-management theory and cross scale thinking)*
Important role of key people at all levels (community, regional, national, global)	All cases but see the Nepal case for a good e.g., or backyard hatcheries in Thailand	The role of leadership and risk taking needs better understanding around evolving extension/ dissemination and related strategies. For instance, *identification of those who can afford (best positioned) to take risks at different stages of change cycles that one may need to target is crucial*

(continued)

Table 10.1 (continued)

Generalized lessons learnt	Example case study(ies)	Implications for future planning[a]
A changed set of incentives	In Vietnam the Doi moi ("opening up") macroeconomic policy played a major role in preparing the policy ground for the rapid expansion of catfish culture	*The Viet Nam case is truly without historical precedence. This rapid change is likely to happen again. What cases might be next and what can be done in preparation?*
Co-production of knowledge among partners and then its packaging, sharing, and adaptation	Backyard hatcheries in Thailand, CBF in Sri Lanka	*Knowledge and its uptake by various audiences need more careful examination including issues such as packaging for dissemination and adaptation*
Adequate resources, particularly financial and human capacities	All	External (often donor) funding/resources have been critically important to provide more flexible resources to bring about change; it also seems to provide increased internal recognition and increases the probability of change
Success equals resilience or persistence over time	Sri Lanka, Nepal, and Thailand maybe best illustrated in the long history of rice-fish systems China	Most cases have continued for 20 years or more (or several thousand years as in the integrated farming in China); *resilience theory could be tested in follow-up research on such questions*
Good science was a critical feature in many successes. (maybe not all; e.g., in rice-fish China)	GIFT case and Catfish seed production in Vietnam; CBF in SL (where choice of reservoir, area for stocking, species densities, combinations etc. were all essential to success)	*Effective processes linking scientists to govt and farmers were essential. Again this should be examined through the opportunity lens and the implication is that these parties need to be part of various dialogue processes*
Organizational changes often took place as part of the overall change process	Indian and Sri Lankan cases	NACSA in India and NAQDA in Sri Lanka were govt responses to some of these changes (opportunities) and represent one change used by various govt agencies. In both instances, the *organizational changes, at the correct juncture, have proved to facilitate success*
Success is contagious!	Adoption of BMPs in shrimp farming in India is perhaps the best example of this trend. However, it is thought that this also is another more widespread measure of success. If it is "good others want it and it spreads"	Successes can be used to identify other opportunities to accelerate change elsewhere. For example, the Shrimp farmer group concept developed in India, was later applied in Aceh, Indonesia. Given the uneven level of aquaculture development in the region, there still is considerable scope for expanded sharing and joint learning, while facilitating further private sector involvement needs more exploration

[a]For example, definition of practices that bear thinking about in the future and their possible use to speed up adaptation

to the latter and in the context of food production systems at large. Food production demands will have to increase in the future, but equally production systems will have to face a more intensified level of competition among each other for natural resources needs; all of these may push these systems into new dimensions of sustainable development, which may well include green house gas emissions as a universal, consumer acceptable measure of efficacy of production.

- Successful developments often seem to have two or more stages: an initial (and often long) period of creating an "enabling environment" followed by rapid expansion, once conditions (acquisition of the required knowledge base, recognition and engagement of the appropriate socio-economic *milieu*, marketing aspects, relevant policy changes) become conducive. Part of the aim of the current exercise is to help others shorten the initial "gestation stage," and we foresee a number of new opportunities emerging in Asia, such as in deltaic regions (e.g., the Irrawady Delta, Myanmar).
- The need to improve communication between sectors and levels of stakeholders: different specialist groups often speak a different "language," do not partner, or maybe are openly antagonistic. To attain success, and therefore, sustainability in the sector, there will be a need to facilitate and evolve more effective communication systems among stakeholders, including planners and policy makers; drawing the latter groups into the early stages of discussion and development will hopefully shorten the process considerably and make it more transparent and effective. Perhaps a "Wikipedia" approach to aquaculture is needed!
- Role of social cohesion and farmer organizations – groups have often failed, particularly when these have been imposed "artificially" by external agencies. Groups are more successful when they form "organically or self organize" among people with common interests and/or problems, and/or for a common purpose: the old adage "a bottom up approach" is often more relevant and efficient than a "top down" approach. Such an approach leads to the following:
    - Good governance required to sustain groups
    - Empowering individual farmers as well as the groups
    - Top is also important – to respond to and create policy change in favor of grass roots
- Developments of robust but simple and easily transferable/adoptable production systems based on appropriate levels of technology that permit flexibility and adaptation to change are ingredients for success. Flexible production systems such as rice-fish culture allow farmers to adapt to changing socio-economic *milieux*, but continue to remain economically viable.
- New technology continues to be an important driver of success. The Vietnam catfish case study demonstrates this well as this production system exploded only after the development of the seed production (breeding and larval rearing) systems. A common factor in most of the case studies is that they were preceded or enabled by a research input (either through public or private sector). However, it needs to be noted that initial development of aquaculture systems is often driven by farmer innovations, which however, is seldom acknowledged.

- Institutes (universities, research institutions, and national, regional and international aquaculture organizations) are clearly important contributors and a part of success. Ways and means of improving these linkages remain an important issue for the future.
- Establishing linkages between aquaculture and other sectors within the broader sustainability framework, for example, who may control policy or access to essential resources, and an understanding of these elements contribute to the attainment of success. This cross-sectoral integration will become increasingly important with time as the envisaged and overly acceptable ecosystem approach to food production systems comes into being.
- Communications, networking, and partnering more generally seem to be a gap in many areas. With globalization, for example, in actively promoting change, in market and value chains, an understanding and compliance to environmental and food quality regulations are becoming increasingly crucial to achieving and sustaining success.
- Key people in various positions, in the private sector and in government, are crucial to facilitate initial breakthroughs and/or farmer innovation that essentially help carry those to the next step.
- Understanding opportunity: if change is on the increase, more work in guiding risk takers, early adopters and individuals who can see opportunities to promote/take advantage of impending changes, so others could follow.
- Establishing a new partnership or regional community for engagement in improving the concepts and dissemination packages on better practices (perhaps a new version of BMPs) to continue to examine "success" and how best to move ahead in this thread of thinking. Part of this new partnership might involve the development of a modified analytical framework for "sustainable success."
  - Situation appraisal: e.g., ability to judge which nonperforming trials should be dropped
  - Development of a wider set of appropriate technologies
  - Dissemination and its newer forms of knowledge sharing
  - Development of a set of longer term indicators for monitoring and evaluation of change and moves toward sustainability, and
  - Creating the right environment where innovators can work more effectively together to share and create change
- Biodiversity conflicts and tradeoffs around use of exotic species and Invasive Alien Species (IAS) have to be addressed on a case by case basis in preference to a broad brush approach, and specific quantifiable indices need to be developed with universal applicability.
- Enhancing capacity on co-management approaches for increasing the efficacy and sustainability (e.g., of CBF) is needed.
- More social analysis research on developing a better understanding of success and group characteristics (social strata, cohesion, economic incentives….).

We are enthusiastic about the start that we have made in a new and challenging area of work, and we look forward to continuing this process using some of the ideas dealt with here. We will continue to reach out to others as part of our plans for continuation of this effort, with a focus on an expanded dialogue with interested stakeholders as part of our efforts of better understanding of how change takes place in this sector and the development of strengthened influencing strategies.

References

Aldhous, P. 2004. Fish farms still ravage the sea. Sustainable aquaculture takes one step forward, two steps back. *Nature Online*, 17 February 2004. doi: 10.1038/news040216-10

Aquaculture Development Beyond. 2000. *The Bangkok declaration and strategy*, 23. NACA/FAO

De Silva, S.S. and G.M. Turchini. 2008. Towards understanding the impacts of the pet food industry on world fish and seafood supplies. *Journal of Agriculture and Environmental Ethics* 21: 459–467

De Silva, S.S., R.P. Subasinghe, D.M. Bartley, and A. Lowther. 2004. Tilapias as alien aquatics in Asia and the Pacific: A review. *FAO Fisheries Technical Paper* 453: 65 pp.

Powell, K. 2003. Fish farming: Eat your veg. *Nature* 426: 378–379

Contributors

Belen O. Acosta, BSc Fisheries (Aquaculture), MSc Fisheries (Aquaculture)
Ms. Belen Acosta is presently the Special Departmental Coordinator for Aquaculture Department of the Southeast Asian Fisheries Development Center (SEAFDEC/AQD). Prior to joining SEAFDEC/AQD, she was a research fellow at WorldFish Center's Aquaculture and Genetic Improvement and assisted in coordinating the International Network on Genetics in Aquaculture (INGA) – a global forum for collaborative research and training in applied fish breeding and genetics. She also led and managed the IDRC-funded collaborative project on "Public and Private Partnerships in Fish Genetics Research" using the GIFT (Genetic Improvement of Farmed Tilapia) experience as an example. Prior to this, she was a research associate at ICLARM's (formerly WorldFish Center) GIFT Project. One of her major contributions in this project was leading the development of the initial version of GIFT Manual of Procedures using the United Nations' standardized training methodology.
SEAFDEC Aquaculture Department, Manila Office, 102 Philippine Social Science Center Building, Commonwealth Avenue, Diliman, 1101, Quezon City, Philippines, bacosta@seafdec.org.ph

Upali S. Amarasinghe, BSc (Kelaniya), MSc (Wales), PhD (Ruhuna)
Upali has 29 years of experience in reservoir fisheries and fisheries management in Sri Lanka. He worked as a research officer in the Ministry of Fisheries and National Aquatic Resources Research & Development Agency (NARA) from 1980 to 1992. Since 1992, he is working as an academic staff member in the University of Kelaniya, Sri Lanka. He worked as a commonwealth postdoctoral fellow in Imperial College, London in 2000–2001 and was an academic visitor in the Centre for Limnology, Netherlands Institute of Ecology, Nieuwersluis, The Netherlands in 2007. He participated in two major research projects on reservoir and lake fisheries in Asia funded by European Commission and Australian Centre for International Agricultural Research. He is the winner of Third World Academy of Science award for the young scientist in the field of Biology in 1993 for research on fisheries biology. Besides winning several awards from the Presidential Award Scheme of Sri Lanka for publishing papers in journals of high impact factor, he also won the Vice Chancellor's award for the outstanding scientist in the Faculty of Science, University of Kelaniya in 2007 and the Committee of Vice Chancellors and Directors (CVCD)

Excellence Award for the most outstanding researcher in 2008. Besides, publishing over 90 papers, he has been a contributing author to a chapter in *Water for Food, Water for Life* in the Global Synthesis, *Comprehensive Assessment of Water Management in Agriculture* (IWMI). He has co-edited the book *Aquatic Ecosystem and Development: Comparative Asian Perspectives*, Backhuys Publishers, Leiden, The Netherlands. He is a Chartered Biologist in the Institute of Biology, Sri Lanka and a life member of several scientific professional societies.
Department of Zoology, University of Kelaniya, Kelaniya GQ 11600, Sri Lanka, zoousa@kln.ac.lk

F. Brian Davy, BSc, PhD
Brian received his PhD in Biology from Texas A&M, focusing on fisheries biology/natural resource management. He has worked as a Postdoctoral Fellow at Cornell University and as a visiting scientist at Tokyo University of Fisheries. Much of his experience has been with IDRC based in Canada (Vancouver and Ottawa), as well as 13 years based in Singapore. His interests lie in sustainable aquatic resources management, particularly in aquaculture and small scale fisheries.
Senior Fellow IISD, 340 Albert St, Suite 710, Ottawa ON, Canada K1R 7Y6, bdavy@iisd.ca

Sena S. De Silva, BSc (Hons), PhD (Stirling), DSc (Stirling), Hon. DSc (Ruhuna U.)
With over 35 years of experience in the academia, and aquaculture and inland fisheries management research and development, he has held academic positions in universities in Sri Lanka, Stirling, Scotland, National University of Singapore, and Deakin University, Australia. He was responsible for developing and delivering postgraduate courses in aquaculture in the "distance mode." He is an internationally reputed researcher in fin fish nutrition and reservoir fisheries, and has expertise in fish introduction and biodiversity in relation to aquaculture. He is the author of three advanced texts and over 200 research publications in international journals and serves on the editorial board of the journal *Aquaculture International, Aquaculture Research, Fisheries Management and Ecology*. He has received many awards including the NAGA Award (ICLARM) in 1993, Deakin University Vice Chancellor's award for "Best Researcher," Asian Fisheries Society Gold Medal in 2004 and Honorary Life Member of the World Aquaculture Society (2005). He was a founder member of the Asian Fisheries Society and served in the Council for 9 years.
NACA, PO Box 1040, Kasetsart Campus, Bangkok 10903, Thailand, sena.desilva@enaca.org

Modadugu V. Gupta, PhD, Hon. DSc
Dr. Modadugu V. Gupta, the former Assistant Director General of WorldFish Center, has over 40 years of experience in fisheries and aquaculture research and development, networking, and human resources development in Asia, Pacific, and Africa. He led and provided guidance to WorldFish Center in determining how its strategic research thrusts could address the country and regional priorities and

become more relevant to the country's developmental needs. He coordinated and managed the International Network on Genetics in Aquaculture (INGA), a global forum aimed at increasing aquaculture production in developing countries in Asia, Africa and Pacific through genetics, breeding, and biodiversity research.
Dr. Gupta has been recognized internationally for his significant contributions in improving the aquaculture sector and in enhancing its greatest potential to contribute to the food and nutritional security of the poor in many developing countries. Apart from helping hundreds of thousands of poor farmers (including women) gain access to fish, he was instrumental in enhancing the awareness of these countries on the importance of biodiversity protocols and management policies that would bring about sustainable fisheries and aquaculture. In view of this, Dr. Gupta was awarded the 2005 World Food Prize – the world's most prestigious award and considered as Nobel Prize in Food and Agriculture.
C502, Aditya Elite, B.S. Maktha, Hyderabad 500016, India, guptamo2000@yahoo.co.in

Tek B. Gurung, BSc, MSc (Gorakhpur U.), DSc (Kyoto Uni)
He has expertise in fisheries and aquaculture in lakes, carp, catfish, rainbow trout farming, and has involved in mahseer (*Tor putitora*) breeding nursing and their aquaculture, management, aquatic ecology & fish biodiversity, and lake conservation and management. He has over 100 publications in international and national journals, proceedings, newsletters, etc.
Fisheries Research Division, Godawari, PO Box 13342, Kathmandu, Nepal, frdgdr@wlink.com.np

Hassanai Kongkeo, BSc, MSc (Microbiology), MSc (Chemical Eng.)
Hassanai was the former coordinator (1996–2001) of NACA. He has worked in the private sector in many capacities, among which are as Managing Director, Saha Marine Products Co., Ltd., Thailand, which was the first *P. monodon* farm in Thailand; Technical Director Chantaburi Farm Co., Ltd., Thailand on shrimp, grouper and oyster grow-out; and as Managing Director, Bangkok Aquafarm Ltd., on seafood marketing and aquaculture consultancy services. In addition, Hassanai has been the Project Manager, World Bank/CDC Aquaculture Project, Thailand that dealt on shrimp, freshwater prawn and seabass hatchery and grow-out, and in the FAO/UNDP/NACA Regional Seafarming Development and Demonstration Project, Thailand on network of seafarming centers in Asia-Pacific. He also has experience in working on aquaculture in Africa and is consulted widely in Asia and elsewhere. He is a technical assistant to the Director General, Network of Aquaculture Centres in Asia-Pacific.
PO Box 1040, Kasetsart Post Office, Bangkok 10903, Thailand, hassanai.kongkeo@enaca.org

Jiashou Liu, BAq, MSc, PhD
Liu Jiashou is a research professor of the Institute of Hydrobiology, Chinese Academy of Sciences, and also the deputy director of the National Research Centre

for Freshwater Fisheries Engineering of China. He has served as council members of several professional societies. He has been involved in a number of national and international research projects in different disciplines, and has also acted as consultant to international and local projects conducted in China. His research interest covers lake and reservoir fisheries management, aquaculture, and fisheries ecology, with focus on cage culture, lake and reservoir enhancement, and freshwater fisheries management. He has built up wide collaborative relationships with many foreign scientists, including those from Czech Republic, Denmark, Canada, the U.S, Australia, Thailand, etc. He has published over 50 research papers and two professional books.
Institute of Hydrobiology, Chinese Academy of Sciences, 7 South Donghu Road, Wuhan 430072, PR China, jsliu@ihb.ac.cn

Miao Weimin, Master of Aquaculture
Mr. Miao Weimin joined Freshwater Fisheries Research Centre in 1982. He has worked as a researcher, lecturer, and administrator ever since he joined the institution. He headed and participated in a number of research projects on carp genetic improvement, inland open water fisheries and socioeconomics, etc. He is also an experienced lecturer and postgraduate supervisor of Nanjing Agricultural University. He worked as UNDP consultant under HDI Program in Myanmar in 1995, where he successfully set up a number of pioneer aquaculture operations in hilling Shan State. As the coordinator, he successfully organized nearly 30 international training courses/workshops on aquaculture and sustainable inland fisheries development, which were attended by some 800 fisheries officers from more than 70 countries in the past two decades.
Miao has a rather broad knowledge and practical working experiences on aquaculture and inland fisheries development. He has good understanding on prospects and development strategies of aquaculture and inland fisheries in the region. He is currently the vice chair of Specialized committee on Inland Aquaculture of China Fishery Society and Board Member of Asia-Pacific Chapter of World Aquaculture Society.
Mr. Miao Weimin was appointed by FAO as the aquaculture officer at the FAO Regional Office for Asia and the Pacific in July 2008.
FAO Regional Office for Asia and the Pacific, Maliwan Mansion, 39 Phra Atit Road, Bangkok 10200, Thailand, Weimin.Miao@fao.org

Nguyen Thanh Phuong, BSc, MSc (AIT), PhD (Toulouse)
Phuong has more than 20 years experience in aquaculture research, teaching, and development. He has been involved in a number of projects in different disciplines and has also acted as consultant to international and local projects conducted in Viet Nam, especially in the Mekong Delta. Phuong has, for example, worked on the development of feed and feeding for catfishes (*Pangasius bocourti* and *Pangasianodon hypophthalmus*) and indigenous fishes, development of better management practices (BMP) for Pangasius catfish production in Viet Nam, development of analytical and biological methods in support of sustainable aquaculture practices in Viet Nam, targeting catfish and shrimp (2007–2009), training and

research on physiological constraints in aquaculture in the Mekong targeting on catfish and black tiger shrimp (2007–2009), and on diseases and water quality in catfish production systems (2004–2007). He is the author/co-author of more than 60 publications in national and international peer reviewed journals, conference proceedings, and book chapters.
College of Aquaculture and Fisheries, Can Tho University, Campus 2, 3/2 Street, Ninh Kieu District, Can Tho City, Vietnam, ntphuong@ctu.edu.vn

Thuy T.T. Nguyen, B.Sc (Nha Trang), Masters in Aquaculture (Deakin), PhD (Deakin)
Thuy is highly experienced in application of genetics in aquaculture and fisheries management. She has a wide and up-to-date knowledge on population genetics, phylogenetics, conservation genetics, and applications thereof. She has been involved in a number of projects in the Asia-Pacific region pertaining to application of molecular genetics to broodstock management and conservation of aquatic species (e.g., in the case of mahseer species in Sarawak, Malaysia and the critically endangered Mekong giant catfish), capacity building in the use of genetic tools in aquaculture and fishery management. Thuy is experienced in organizing training courses as part of the work program at NACA. She conducted two successful hand-on training courses in application of molecular genetic tools in aquaculture and inland fisheries management, in conjunction with the Food and Agriculture Organisation (FAO), Kasetsart University and Department of Fisheries Thailand, and the Central Institute of Freshwater Aquaculture, India, and a training course on conservation genetics in conjunction with the University Sains Malaysia. Thuy has authored/co-authored 25 publications in peer-reviewed international journals, two manuals, one book chapter, a number of popular articles, and edited one book. Thuy is co-supervising a number of students undertaking postgraduate studies. Thuy was a recipient of the best young agricultural researcher from developing countries award for 2008 of the Japanese Ministry of Agriculture, Forestry and Fisheries.
PO Box 1040, Kasetsart Post Office, Bangkok 10903, Thailand, thuy.nguyen@enaca.org

P.A. Padiyar, BFSc, MFSc
Arun is an aquaculture specialist and is a consultant to FAO and World Bank Group, and presently, he is a pursuing PhD at Deakin University, Australia on a part-time basis. During the last 9 years of his professional career, he worked on aquatic epidemiology, extension, better management practices, cluster farm management in small-scale shrimp farming, and development of rural livelihoods. He is also specialized in postdisaster and postconflict aquaculture rehabilitation and development activities. During 2001–2005, he worked as Project Supervisor for MPEDA-NACA project in India. Since August 2005, he is working in tsunami affected Aceh, Indonesia.
Jln Angsa No. 19, Desa Ateuk Deah Tanoh, Banda Aceh, 23244, Nanggroe Aceh Darussalam, Indonesia, arunpadiyar@gmail.com

Michael J. Phillips, BSc, PhD

Dr Michael Phillips obtained his PhD in Aquaculture from the University of Stirling, Scotland in 1982. He has extensive experience with aquaculture in Asia, working for FAO, the Asian Institute of Technology and, since 1994, with the Network of Aquaculture Centres in Asia-Pacific (NACA). In his position as Manager of Research and Development Programs and Environment Specialist at NACA, Dr Phillips manages a wide range of aquaculture projects and development initiatives throughout the Asian region, including pioneering work on environmental management of shrimp farming, which received a World Bank Green Award in 2006. He has been actively involved in the MPEDA/NACA project since 2000.

PO Box 1040, Kasetsart Post Office, Bangkok 10903, Thailand, Michael.phillips@enaca.org

Ash K. Rai, MSc (T.U.), MSc (Auburn U.), PhD (Kyoto U.)

Rai has over 30 years of experience in inland aquaculture development and research. As head of the Fisheries Research Organization, Agricultural Research Council of Nepal, he was responsible for planning, programming, and implementation of research and development activities, launched different projects supported by different international donors, and acted as consultant for carp culture development in Papua New Guinea, Rainbow trout culture in Chengmai, Thailand, aquatic ecologist for ecological studies and impact assessment, aquaculture consultants for small scale farmers, and short-term FAO aquaculture consultant. Rai is the recipient of the Mahendra Bidhyabhusan Padak (Ka), RONAST Science and Technology Promotion Award for the contribution to fisheries research and development in Nepal, and has published over 80 papers in national and international journals. He has membership in several scientific professional societies including Asian Fisheries Society and Japan Limnological Society.

Resources Himalaya Foundation, Conservation Chautari, Naya Bato, Lalitpur, GPO Box 2448, Kathmandu, Nepal, akrai@resourceshimalaya.org, www.resourceshimalaya.org

Umesh, N.R., BFSc, MFSC

Umesh is a Fisheries Graduate from College of Fisheries, Mangalore, with 15 years of experience in Shrimp and Tilapia Aquaculture projects in India, Jordan, and Ghana. He also worked as Project Supervisor in the MPEDA-NACA village demonstration program, which was a collaborative project between Marine Products Export Development Authority (MPEDA), Govt. of India and Network of Aquaculture Centres in Asia Pacific (NACA), Bangkok, Thailand on shrimp disease control in India.

National Center for Sustainable Aquaculture, 69-17-8, SBI Officers Colony, Rajendra Nagar, Kakinada-533003, AP, India, nacsa.hq@gmail.com

Index

The manufacturer's authorised representative in the EU is Springer Nature Customer Service Centre GmbH, Europaplatz 3, 69115 Heidelberg, Germany. If you have any concerns regarding our products, please contact ProductSafety@springernature.com

Printed and bound by CPI Group (UK) Ltd, Croydon, CR0 4YY
30/09/2026
02225510-0004